Geometry of Meaning

Geometry of Meaning

Semantics Based on Conceptual Spaces

Peter Gärdenfors

The MIT Press
Cambridge, Massachusetts
London, England

MIT Press books may be purchased at special quantity discounts for business or sales promotional use. For information, please email special_sales@mitpress.mit.edu.

This book was set in Stone Sans and Stone Serif by Toppan Best-set Premedia Limited, Hong Kong. Printed and bound in the United States of America.

Library of Congress Cataloging-in-Publication Data

Gärdenfors, Peter.
Geometry of meaning : semantics based on conceptual spaces / Peter Gärdenfors.
    pages cm
Includes bibliographical references and index.
ISBN 978-0-262-02678-9 (hardcover : alk. paper) 1. Semantics—Psychological aspects. 2. Cognition. 3. Cognitive science. 4. Psycholinguistics. I. Title.
P325.5.P78G35   2014
401'.43–dc23
2013022298

10 9 8 7 6 5 4 3 2 1

Contents

9   Events   159

10   The Semantics of Verbs   181

11   The Geometry of Prepositional Meaning   201

Preface

My ambition in this book is to present a cognitive theory of semantics that unifies a large number of areas. It is a continuation and development of the ideas in my book *Conceptual Spaces: The Geometry of Thought.*

The savanna of semantic research is populated by different kinds of actors. I belong to the giraffes, who try to get a view of the entire landscape. My goal is to present, if not the geography of the meaning space, at least its geometry. Another species of researchers is that of the dung beetles, often found in linguistics departments, who collect their piece of language, for example, a syntactic oddity, and turn it over and over again until they have extracted all its contents. The problem for the giraffe is to be able to reach down to the empirical ground and grasp all the interesting details, while the problem for the dung beetles is to see how the landscape of meaning hangs together. As is well known, these species, and many others, are needed for a well-functioning research landscape. Although my goal is to present a general view of semantics, I hope the book will generate a lot of material for the dung beetles to work on. If my theses are valid, at least to a first approximation, they should provide a rich source for further investigation by researchers in linguistics, cognitive science, and developmental psychology. If not, I would like to see my theses refuted.

I have not been alone in developing the theory presented in this book. Above all, I want to thank my friend and collaborator Massimo Warglien. We have worked together on several of the topics presented here, and we have written a series of papers that have been transformed into chapters or parts of chapters. He is coauthor of most of the material in chapter 4, sections 5.4–5.7, chapter 9, chapter 10, sections 13.1–13.3, and the appendix. He has also drawn several of the figures in the book. During our stay in Uppsala during the fall of 2012, Joost Zwarts was a rich source for me concerning linguistic research related to the topics of the book. In particular, we considered the semantics of prepositions, a discussion that resulted in a joint paper that basically contains the material in sections 11.2–11.6 (the two texts were written more or less in parallel). I also want to thank Simone Löhndorf as a coauthor of a paper that contains the material covered in sections 2.5 and 2.6, and Matthijs Westera, who is a coauthor with Massimo Warglien and me on a paper about events and the semantics of verbs that contains large parts of what is presented in chapter 10 and some parts of chapter 9. Consequently when I write "I," I often mean "we," although the "we" is different in different parts of the book. The texts of the coauthored articles have been modified, though, to fit the book as a whole, and I am solely responsible for the results.

Apart from these persons, I have received helpful comments from Richard Andersson, Manfred Bierwisch, Ingar Brinck, William Croft, Markus Egg, Elisabeth Engberg-Pedersen, Hans-Martin Gärtner, Emmanuel Genot, Marianne Gullberg, Paul Hemeren, Jim Hurford, Ray Jackendoff, Justine Jacot, Gerhard Jäger, Ingvar Johansson, Martin Jönsson, Dunja Jutronic, Manfred Krifka, Werner Kuhn, Ronald Langacker, Asifa Majid, Sabine Marienberg, Carita Paradis, Eric Pedersen, Oleksiy Polunin, Uli Sauerland, Chris Sinha, Leonard Talmy, Jürgen Trabant, Susanne Vejdemo, Annika Wallin, Jordan Zlatev, and participants in the CCL (Cognition, Communication, and Learning) seminars during the spring semester of 2012, when I presented a first draft of the book. I also want to thank Rasmus Bååth and Magnus Haake for their help in creating the illustrations in the book and Alan Crozier for checking my English.

During the period of writing, I was generously supported by a number of institutions. I gratefully acknowledge the Swedish Research Council for support to the Linnaeus environment Thinking in Time: Cognition, Communication, and Learning (CCL). My home base, LUCS (Lund University Cognitive Science), and the rest of the Department of Philosophy have provided a critical but friendly test bed for my ideas. Second, the CCL environment in Lund has generated opportunities for new collaborations that have been useful for my research. During the fall semester of 2011, I spent a rewarding period at the Forscherkolleg Bildakt und Verkörperung at the Humboldt Universität in Berlin, where I had many fruitful discussions concerning visual processes, embodiment, and meaning. During my Berlin sojourn I also had valuable discussions with linguists, in particular Manfred Krifka, at the Zentrum für Allgemeine Sprachwissenschaft (ZAS). Then, during the fall of 2012, the Swedish Collegium of Advanced Studies (SCAS) in Uppsala provided a highly creative milieu for me and a group of other semanticists.

I SEMANTICS AS MEETINGS OF MINDS

1 What Is Semantics?

1.1 What Does It Mean to Understand Communication?

What is it that you know when you know a language? Certainly you know many words of the language (its lexicon), and you know how to put the words together in an appropriate way (its syntax). More important, you know the *meaning* of the words and what they mean when put together into sentences. In other words, you know the semantics of the language. If you do not master the meaning of the words you are using, there is no point in knowing the syntax. Therefore, regarding communication, semantic knowledge is more fundamental than syntactic.[1]

You can communicate in a foreign language, with some success, just by knowing some words and without using any grammar. Or if you do not know any words of the culture you happen to be in, you can try communicating by gestures, hoping that the intended meaning of your gestures will be transparent to your interlocutor. Thus meaningful communication can occur without a common language. It is not only spoken and written language that has meaning, but also other communicative acts. A continuity exists between verbal and nonverbal communication systems in the way meanings are expressed. A theory of semantics should account for this continuity.

De Saussure (1966) distinguishes between *langue* (language as a social system of signs) and *parole* (the individual speech acts). He writes that the signs of language are "collective products of social interaction, essential instruments through which human beings constitute and articulate their world," but he never explains how *langue* can arise from *parole*. This book presents a theory of semantics that comes in three parts. Part I focuses on how various forms of communication establish a system of meanings that becomes *shared* between the interlocutors.[2] This will explain the emergence of *langue*.

Many semantic theories take the meanings of words to be more or less stable and independent of the communicative context. In contrast to how semantics have been studied in much of the logical tradition of philosophy or in mainstream linguistics, I argue that semantics develop as an *interplay* between communicative acts, in particular *speech acts* (Searle, 1969) and already existing meaning conventions. From this perspective, we can draw no sharp boundary between pragmatics and semantics: semantics can be characterized as conventionalized pragmatics (Langacker, 1987, sec. 4.2).

As I explain in greater detail in chapter 5, meaning is often *negotiated* in a conversation. Some meaning changes are quick, such as when a pronoun changes its reference; some are slower, such as when two speakers find out that they use the same word in different senses and have to coordinate; and some are very slow, such as when the meaning of a word changes over historical time. Meanings of words coevolve on these different timescales, and a semantic theory should account for this.

The theory presented in the first part of the book will connect the semantics of various forms of communication to other cognitive processes, in particular concept formation, perception, attention, and memory. As Jackendoff (1983, p. 3) puts it: "To study semantics of natural language *is* to study cognitive psychology."

My theoretical starting point is that our minds organize the information that is involved in these processes in a format that can be modeled in geometric or topological terms—namely, in *conceptual spaces*. I presented the theory of conceptual space in an earlier book (Gärdenfors, 2000). Here I extend it to a more general theory of semantics. After some preparatory chapters, the communicative perspective will be integrated with the geometric in chapter 5. The main result is that "well-shaped" meaning structures will make communication efficient.

Part II focuses on *lexical semantics*. My aim is to show that by using conceptual spaces, we can develop a unified theory of word meanings.[3] A subgoal is to show how the meanings of different word classes can be given a cognitive grounding. I expand on the analysis of nouns and adjectives that I outlined in Gärdenfors (2000). A theory of the semantics of verbs is built on the basis of cognitive theories of actions and events, and one chapter is devoted to prepositions. I also present some models of how the meanings of words are *composed* to form new meanings and what the basic semantic role of sentences is.

Throughout the book, I formulate a number of semantic theses that I believe to be novel. For several of them, my empirical evidence is meager, so they require further investigations to be confirmed. I invite linguists and cognitive scientists to a critical evaluation of the theses. The brief Part III contains some implications of the theory for robot semantics and the Semantic Web.

1.2   Criteria for a Theory of Semantics

When formulating a theory of semantics, one should specify criteria for the success of the theory. In previous work (Gärdenfors, 1997, 2000), I formulated four basic criteria for a theory of semantics.[4]

(1) *The ontological criterion*: A theory of semantics shall explain what kind of entities meanings are.

Within philosophy of language, one can find two fundamentally different answers to the ontological question, one *realistic* and one *cognitive* (or *conceptualistic*). According to the realistic approach to semantics, the meaning of a word or expression is something *in the external world*.[5] However, most of the contents of human communication concern our *inner worlds*: our memories, judgments, plans, fantasies, and dreams. The primary use of language is to refer to objects that are not present in the scene of communication. Hockett (1960) proposes that this kind of "displacement" is one of the main factors that distinguishes symbolic language from mere signaling. Therefore a semantic theory for communication that builds only on reference to the external world seems unnatural.

According to the cognitive tradition, meanings are *mental entities*. The prime slogan for cognitive semantics has been *meanings are in the head*. More precisely, a semantics for a language is seen as a relation between the communicative acts and some cognitive entities. Later in the chapter, I describe some theories about how these cognitive entities are constituted. Note, however, that a cognitive approach to semantics does not entail that the external world plays no role in determining the contents of the meanings in the head. On the contrary, our cognitive structures are formed in constant interplay between our minds and the external world.

Depending on which account one chooses for the ontological criterion, the second criterion will be handled in two radically different ways.

(2) *The semantic criterion*: A theory of semantics shall account for the relation between communicative expressions and their meanings.[6]

The realist interpretation of the semantic criterion is often formulated in terms of *truth conditions* based on a mapping between elements of a language and the external world. This approach to semantics brackets the role of humans (and other potential language users) in the connection between a communicative system and its meaning.[7] In contrast, a cognitive semantics puts the role of humans in the center. Communication is seen as one among many cognitive activities and as tightly interwoven with these.[8]

The two traditions have different ways of identifying the central semantic relation. Realist semantics has focused on sentences, while the meanings of the building blocks (individuals and predicates) have been taken as given. In contrast, cognitive approaches have mainly dealt with the meanings of single words and simple expressions.[9] A special aspect of the semantic question is how meanings of *composed expressions* relate to the meanings of the constituents. This criterion is what Jackendoff (1983, p. 11) calls the compositionality constraint. In chapter 13, I explain how the semantic theory of this book handles compositionality.

It is commonplace that language is conventional. Consequently the semantics of a language have to be *learned* by individual speakers. This leads to the third criterion:

(3) *The epistemological criterion*: A theory of semantics shall explain how the meanings of communicative acts can be learned.[10]

Children learn a language without effort and completely voluntarily. They learn new words miraculously fast. A teenager masters about sixty thousand words of her mother tongue by the time she finishes high school. In her speech and writing, she may not actively use more than a limited subset of the words, but she *understands* them. A simple calculation reveals that she has learned an average of nine or ten words *per day* during her childhood (Carey, 1978). A single example of how a word is used is often sufficient for learning its meaning.[11] No other form of learning is so efficient. Yet the underlying learning mechanism is still to a large extent unknown.

How do children know what to learn? When a new word is uttered in a situation, there is often a multitude of objects, features of objects, and features of the ongoing events that could be the meaning of the word. How does the child select the right meaning?[12] Several cues can help the child. A first cue is that words do not mean just anything. Our minds perceive different kinds of structures in the world, and the meanings, to a large extent, follow these structures. Above all, there are *things* in the world that we need to talk about. The things come in classes, and we have words for them that have meanings that correspond to these classes. Another cue is that the child's *attention* is often focused on a single object (Tomasello, 2001). Yu and Smith (2012, p. 3) argue that "the premise of referential ambiguity is exaggerated. Early word learning often takes place in the context of infants' active exploration of objects: infants do not simply look passively at the jumble of toys on the floor but rather use their body—head, hands, and eyes—to select and potentially visually isolate objects of interest, thereby reducing ambiguity at the sensory level." A third type of cue is that the syntactic markers of a new word often reveal its word class (Bloom, 2000, chap. 8). The markers help the child to look for an object when a noun is used, a property of an object when an adjective is used, an aspect of the event when a verb is used, and so on. A fourth type of cue is the feedback given when the child tries out a new word.

For these reasons, a semantic theory should generate plausible constraints on language acquisition. In particular, the constitution and function of human short- and long-term memory should be compatible with the semantic structures. I formulate two more criteria connecting cognitive processes to semantics later in the chapter.

Even if we assume that all speakers use the same vocabulary and grammar and that each speaker is able to determine what is meant by the linguistic expressions, it remains to be explained how they can mean the *same* thing. Therefore we should endeavor to explain how individual semantic mappings are *coordinated* in communication. This leads to the fourth criterion:

(4) *The social criterion*: A semantic theory shall account for the relation between the meaning systems of individual speakers and their communal language.[13]

The social criterion concerns how we can assume that we are talking about the same things.[14] Realist theories of semantics have no problem with this criterion: since the meanings of the words are out there in the world, we will talk about these things as soon as we speak the same language, that is, as soon as we share the mapping from linguistic expressions to the world. The question is then reduced to the epistemological one, that is, how we learn the mapping. On the other hand, the social criterion is a genuine problem for cognitive theories of semantics.

To these four criteria, which are discussed further in Gärdenfors (2000), I now add two more concerning the connections between semantics and different *cognitive processes*:

(5) A semantic theory shall account for the relation between *perceptual processes* and meaning.
(6) A semantic theory shall account for the relation between *actions* and meaning.

A good justification for adding criteria 5 and 6 comes from Regier (1996, p. 27):[15]

The idea is that since the acquisition and use of language rest on an experiential basis, and since experience of the world is filtered through extralinguistic faculties such as perception and memory, language will of necessity be influenced by such faculties. We can therefore expect the nature of human perceptual and cognitive systems to be of significant relevance to the study of language itself. One of the primary tasks of cognitive linguistics is the ferreting out of links between language and the rest of human cognition.

In the following two sections, I introduce criteria 5 and 6 in greater detail.

1.3   The Relation between Perception and Semantics

1.3.1   The Translation Problem

How can we talk about what we see? This is basically a question of translating between two different types of information: the visual and the linguistic (Macnamara, 1978; Jackendoff, 1987b, 2012).[16]

Signals from the retina are processed in several stages as they spread through the brain, and they are transformed into a visual representation that we do not fully understand, though researchers have made much progress during the last decades. Similarly, language is represented on several levels by the brain. Somehow these types of representation must be intertranslatable, since we can effortlessly talk about what we see; and conversely, when we hear someone tell a story, we immediately form a vivid inner image of the narrative (and we can, more or less successfully, transform this image into a drawing or a sketch).

I have no ambition to solve the translation problem here. However, I want to call attention to some deep correspondences between perceptual processes and semantic mechanisms. I focus on vision, but similar correspondences hold also for the other senses.

1.3.2   Image Schemas in Cognitive Linguistics

Il mondo ha la struttura del linguaggio e il linguaggio ha le forme della mente.[17]
—Eugenio Montale

The core idea of cognitive linguistics is that meanings of linguistic expressions and other communicative acts are *mental entities*. For example, Jackendoff (1983) distinguishes between a real world and a projected world. We have conscious access only to the projected world, which is "the world as unconsciously organized by the mind." Hence, for Jackendoff (1983, p. 29), a clear difference separates reality and conceptual reality: "We can talk about things only insofar as they have achieved mental representation through these processes of organization. Hence *the information conveyed by language must be about the projected world.*"[18]

Most of the analyses within cognitive linguistics concern relations between words and representations of concepts. Central concepts are the *image schemas* developed by Langacker (1987), Lakoff (1987), Talmy (1988), and others.[19] The very word "image" indicates a connection to vision. Every word, possibly except for syntactic markers, is supposed to correspond to an image schema (or several, if the word has several meanings). For example, a verb is represented as a *process* in time. In a simplified way, such a process can be described as involving three stages: one schema part for the beginning, one part for the middle, and one part for the end of the process. A basic image schema for "climb," for example, is represented by Langacker (1987, p. 311), as in figure 1.1.

In this diagram, the time axis is drawn below the three stages. It is marked as a thick line, since the temporal aspect of "climb" is in focus. The *landmark* is supposed to be vertically extended, and the *trajector*, the thing doing the climbing (the small circle), is assumed to be in physical contact with the landmark.

The image schemas have clear connections to perceptual processes; in particular, they have an inherent spatial structure. For example, Lakoff (1987) and Johnson (1987) argue that schemas such as *container,*

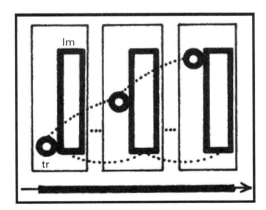

Figure 1.1
Image schema for "climb" (from Langacker, 1987, p. 311). tr = trajector, lm = landmark.

source-path-goal, and *link* are among the most fundamental elements of meaning.[20] In cognitive linguistics, the distinction between trajector and landmark is often claimed to be the same as the distinction between *figure* and *ground* in Gestalt psychology. This analogy is valid for some cases, but not all. In many cases, for example, in the schema in figure 1.1, the landmark is a second object that is related to the trajector, and both function as figures against a general background.

The trajector can be interpreted as the focus of attention. Note that this kind of attention is directed to the "inner space" of representation, not to the ordinary space, as in visual or auditory attention. A fundamental assumption, though, is that the two kinds of attentional processes have essentially the same nature.

According to Langacker, the schema for the verb "climb" can be turned into a schema for "climber" by using the same dimensions, objects, and relations. Only the *focus* of the schema is shifted from the time dimension to the trajector (fig. 1.2).

The transformation can be viewed as an example of *refocusing*. This kind of change has an obvious parallel in vision, where looking at the same scene can generate different cognitive processes depending on which aspects of the scene are focused on.

Holmqvist (1993, p. 31), who develops a formalism for image schemas suitable for computer implementations, connects directly to perception when he defines image schemas as "that part of a picture which remains when all the structure is removed from the picture, except for that which belongs to a single morpheme, a sentence or a piece of text in a linguistic

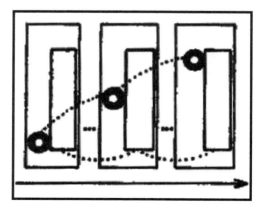

Figure 1.2
Image schema for "climber" (from Langacker, 1987, p. 311).

description of a picture." Mandler (2004, pp. 81–82) relates image schemas to topological representations:

An image of a container . . . must have a particular shape, and the material inside it either conforms to the shape of the container or not, but a topological representation of this relation eliminates this information, leaving only the topological relation of a bounded space with an inside and an outside. In this sense image schemas are topological: They simply do not include some of the information that might be in an image.

The most condensed account comes from Gibbs and Colston (1995, p. 349), who define image schemas as "dynamic analog representations of spatial relations and movements in space." Unlike Lakoff and Langacker, who focus on the spatial structure of image schemas, this definition puts the dynamics of the representations in focus. As we shall see, this aspect is necessary to account for the semantics of verbs.

1.3.3 Parallels between Language Understanding and Visual Processes

After presenting image schemas as semantic tools, I next point out some parallels between semantics, on the one hand, and ordinary visual processes and visualization, on the other. I want to demonstrate that the two kinds of processes share many more similarities than we are normally aware of (also cf. Chafe, 1995; Gärdenfors, 1995, 2004a). My aim is to show that semantics and perception, in particular vision, cannot be separated. As a consequence, the structure of visual processes will *constrain* semantic representations.

Already in the example of the image schemas for "climb" and "climber," we saw an example of *refocusing*, that is, a shift of attention from certain aspects of the schema to others. In cognitive semantics, such attention shifts are described under the name of "reprofiling" (Langacker, 1987; Holmqvist, 1993). I submit that this involves essentially the same process as attention shifts in ordinary vision.[21] The difference is that in the case of image schemas used in language understanding (as well as in inner visualization), the attention changes over an inner image, while in regular vision it moves over the scene presented by the external world.

An interesting case of inner attention is *fictive motion*, which shows up in many linguistic constructions (Talmy, 1996, p. 211; Jackendoff, 2012, p. 1147):

(1.1)   This fence goes from the plateau to the valley.

(1.2)   I looked out past the steeple.

(1.3)   I aimed my camera into the living room.

These constructions do not describe processes that occur in the real world. Instead they represent our understanding of how we "perceive" in our inner world: you follow the fence with your gaze, your gaze moves along the landscape, and the "vision" of the camera is directed to a particular area.

Matlock (2004) argues that fictive motion is manifested as a form of cognitive simulation. She conducted a series of experiments where subjects read stories about travel, for example, fast travel versus slow, and short distance versus long. When the subjects then made a decision about a fictive motion, their response latencies were shorter after having read about fast travel and short distances. Matlock interprets the result as showing that the subjects mentally simulate visual scanning while processing sentences involving fictive motion. This interpretation fits the mental imagery studied by Kosslyn (1980) and others.

This study is an example of a new tradition that emphasizes the role of mental simulation in cognitive processes. Barsalou (1999) argues that concepts should be understood as *perceptual symbols* that are dynamic patterns of neurons functioning as simulators that combine with other processes to create conceptual meaning. However, in Barsalou's theory, these meaning carriers are still closely related to perceptual processes (hence their name).

Fictive motion is also involved in the kind of refocusing that Lakoff (1987) calls "path-process interchange." In a sentence such as "The power line stretches over the yard," the power line does not literally stretch itself (fig. 1.3). However, if the power line is seen as a path that one follows with one's attention, one's mental gaze can be described as "stretching" over the yard. In brief, the path is transformed into a visual process.[22]

A related notion is discussed by Langacker (1987, sec. 3.1.2), who argues that we can perform a *mental scanning* of various forms of mental

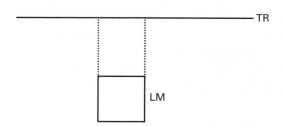

Figure 1.3
Image schema for "The power line stretches over the yard" (from Lakoff, 1987, p. 426). Reprinted with permission University of Chicago Press © 1987.

structures. For example, the following two sentences describe the same scene (Langacker, 2008, p. 82):

(1.4)   The hill gently rises from the bank of the river.

(1.5)   The hill gently falls to the bank of the river.

The difference between (1.4) and (1.5) is the direction of the mental scanning.

He distinguishes between two forms of mental scanning: *sequential scanning*, where one mentally traverses the objects (or path), having only one element active at a time; and *summary scanning*, where the elements are accumulated as we traverse them.[23]

Two other visual processes are *zooming in* and *zooming out*. These processes have clear parallels also in semantics. For example, the location of an object can be described as zooming in (Langacker, 2008, p. 81):

(1.6)   Your camera is in the study, in the closet, on the top shelf, behind some boxes.

The same task can also be achieved by zooming out:

(1.7)   Your camera is behind some boxes, on the top shelf, in the closet, in the study.

However, a mixture of the processes does not work:

(1.8)   *Your camera is on the top shelf, in the study, in the closet, behind some boxes.[24]

Another kind of reprofiling is called *multiplex-mass interchange*. In Lakoff's example

(1.9)   There were soldiers posted all over the hill,

the soldiers are seen not as a multiplex of individual soldiers but as a mass in analogy with

(1.10)   He spilled wine all over the tablecloth.

Visually this corresponds to changing the *scale of resolution*. It can be described as a squinting with your inner eye. The process is not just zooming out; the reprofiling also involves a change in the perception of the soldiers: they are seen as a continuous substance that can be *spread* over the hill.

One of the most general types of refocusing in semantics is *metonymy*. In analogy with the characterization of metaphor in cognitive linguistics (e.g., Lakoff & Johnson, 1980; Lakoff, 1987; Croft, 2002), I regard metonymy as a primarily semantic operation that is expressed in different ways. Classical types of metonymy include *pars pro toto*, where a part of an object is focused instead of a whole object ("there are three sails on the lake," i.e., sails instead of boats), and *totum pro parte*, where the focus is put on something that contains the object as a part ("Paris announces shorter skirts this season," i.e., the city instead of the fashion designers in it). These processes correspond to the process of zooming in (on the sails) and zooming out (to the city of Paris) in visual attention.[25]

A special case of *pars pro toto* is what cognitive linguists call "endpoint focus" (see also Dewell, 1994). For example, Lakoff (1987), in his study of different uses of "over," has looked at cases such as "Sausalito is over the bridge" (from the point of view of San Francisco). Here Sausalito is not literally "over" the Golden Gate Bridge; it is the *path* one travels from San Francisco to Sausalito that goes "over" the bridge (this path is only a visualized one, if we perform a mental scanning of the trip from San Francisco to Sausalito) (fig. 1.4). However, by focusing one's attention on the endpoint of this path, which is Sausalito, one creates a *pars pro toto* metonymy.

These examples suggest that many of the transformations that have been described by cognitive linguists have close analogues in visual processing, in particular focusing of attention, visual scanning, and change of resolution. Nevertheless the analogies require careful experimental testing to determine their validity.

Not only can we talk about what we see, but we also see (and hear, etc.), in our inner worlds, what we talk about. Language and perception are communicating vessels: I regard this as one of the main foundations for semantics. To develop the analysis of the connections between cognitive semantics and perception, we need a principled theory of image schemas and their transformations (for an interesting attempt, see Clausner & Croft,

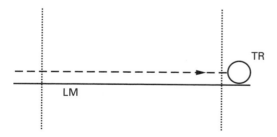

Figure 1.4
Image schema for "Sausalito is over the bridge" (from Lakoff, 1987, p. 422). Reprinted with permission from University of Chicago Press © 1987.

1999). The mathematical structures of image schemas are seldom spelled out.[26] A natural way is to exploit topological and geometric notions. The fruits of such an analysis not only will be useful for lexical semantics but can also be applied in communication models and in pedagogical practices. Furthermore, the analysis can explain the efficacy of metaphors and metonymies. They are not just decorations but powerful tools of communication and education.

I have focused on the visual processes connected with the understanding of single words or simple expressions. However, other areas of language use can also be related to vision. Holsanova and her colleagues have used eye-tracking techniques to systematically investigate the connections between how people *look* at pictures and how they *describe* what they see (Holsanova, 2008; Johansson, Holsanova, & Holmqvist, 2006). For example, one experiment studied how subjects were moving their eyes when retelling the contents of a picture they had previously been looking at. The subjects' eye movements (over an empty white screen) closely corresponded to the eye movements when they first looked at the picture. Furthermore, the grammatical and narrative organization of the story was to a large extent structured in accordance with the first eye-movement patterns. In contrast, if subjects were instructed to keep their eyes fixed on a point when they retold what they had seen, the structure of the reports was different. These results support the thesis that the way we look at the world influences how we talk about what we see. The idea is that the mental representation of the picture has many structural features in common with the real picture. This correspondence has also received support from research on mental imagery (e.g., Kosslyn, 1996).

Although I have focused on vision in this section, it should be emphasized that other senses are also involved in the mental representations of semantic structures. For example, the meaning of a word like "climb" is tied not only to vision but also to tactile perception and kinesthetic experiences, as I argue in section 10.4.

1.4   The Relation between Actions and Semantics

Our talk gets its meaning from the rest of our activities.
—Ludwig Wittgenstein

A main theme of this book is that not only must linguistic information be compatible with the perceptual system, but our *actions* should also be considered. One characteristic feature of actions is that they involve *forces*, typically exerted by an agent. Forces should consequently be among the building blocks of cognitive semantics—not just spatial relations and other perceptual primitives. Language and action are also communicating vessels.

I submit that the domain of forces is as central to semantics as is the spatial domain. In line with this idea, Johnson (1987, p. 42) argues that forces form perceptual gestalts that serve as image schemas (the word "image" may be misleading here, since he connects it not just to visual processes):

Because force is *everywhere*, we tend to take it for granted and to overlook the nature of its operation. We easily forget that our bodies are clusters of forces and that *every* event of which we are a part consists, minimally, of forces in interaction . . . We *do* notice such forces when they are extraordinarily strong, or when they are not balanced off by other forces.

In spite of the interest in embodied cognition, the role of forces and actions is not well developed in cognitive linguistics. For example, Langacker's (1987) analysis of "climb" that I presented earlier does not include forces.[27] A well-known exception is Talmy's (1988) force dynamics. However, his approach provides only a partial model, since it is limited to the interactions of an Agonist (agent) and an Antagonist. One of my aims in this book is to present a cognitively grounded theory of forces and actions.

More recently, one can distinguish a more *dynamic* and *embodied* view of image schemas (which are thus no longer mere "images") (see, e.g., A. Clark, 1997). The focus of the embodied semantics is the body of the agent and its relation to cognitive processes. Johnson (1987) has already written about "the bodily basis of meaning, imagination and reason." Also in Piaget's sensory-motor schemas, which were developed for modeling cognitive development and not for semantics, motor patterns are central.[28] These can be seen as a special case of the dynamic patterns that form our fundamental understanding of the world. However, Mandler (2004, p. 118) writes that "sensorimotor schemas are dynamic structures controlling perception and action, not meanings onto which relational morphemes can be mapped. An interface between sensorimotor activity and its continuously changing dynamics and the discrete propositional system of language is needed." She adds that such an interface should have two characteristics: "It should provide a simplified packaging of preverbal experiences. . . . Their generalizable aspects need to be distilled. . . . Second the interface needs to be in the form onto which a discrete symbol system can be mapped" (Mandler, 2004, pp. 118–119). The conceptual spaces as presented in Gärdenfors (2000) and further developed in this book are presented as such an interface.

In neurolinguistics, Pulvermüller's (2003) results provide important indications of the connections between meanings and actions. He focuses on the *motor* aspects of meaning schemas. When the brain understands a verb, it prepares an action. For example, he has shown that when you read the word "kick," the same part of motor cortex is activated as when you actually kick. An interpretation is that the brain simulates the action it reads about.

In chapter 8, I present an analysis of forces and, more generally, force patterns, and I analyze actions in terms of forces. Then, in chapters 9 and 10, the model of events and the semantic analysis of verbs will to a large extent depend on forces involved in actions.

1.5 Semantics as Meetings of Minds

All speech, written or spoken, is in a dead language until it finds a willing and prepared hearer.
—Robert Louis Stevenson

1.5.1 Meanings Ain't in the Head

A fundamental drawback of the theory of image schemas is that the schemas are supposed to belong to individual minds. The theory of cognitive semantics does not account for how image schemas can be compared between individuals. In other words, mainstream cognitive semantics does not provide a viable account of the social criterion. It seems to be implicitly assumed that all individuals within a language community have the same schemas. However, this assumption would be difficult to combine with any reasonable theory of how the image schemas are learned.

Of course, it can be said that words (or signs) referring to something in one person's inner environment can be used to communicate as soon as the listeners have, or are prepared to add, the corresponding references in their inner environments. The actual conditions of the outer situation need not play any role for the communication to take place: two prisoners can talk fervently about life on a sunny Pacific island in the pitch dark of their cell. But how do we know that other people have "corresponding" references in their inner environments when they use the same words that we do?

Animals use a variety of signs to communicate. Human communication involves not just an understanding of the meaning of a sign itself but an understanding that a sign can be used communicatively. Zlatev, Persson, and Gärdenfors (2005) call this the *communicative sign function* and define it as follows: "The subject intends for the act to stand for some action, object or event for an addressee, and for the addressee to appreciate this."

The communicative sign function implies an understanding that the sign has *the same meaning for the addressee as for the sender*. But "having the same meaning" is a reflexive notion, and this implies at least some degree of third-order intersubjectivity (see chap. 3). Consider the simple example of what knowing the meaning of the word "cat" implies:

(i) I know that "cat" means "a small furry animal that meows."
(ii) I assume you know that "cat" means "a small furry animal that meows."
(iii) I assume that you know that I know that "cat" means "a small furry animal that meows."

The processes of how the inner worlds of the interlocutors are coordinated is the focus of the first part of the book.

A heavy attack against the very possibility of cognitive semantics, with respect to the social criterion, was launched by Putnam (1975). He summarizes his argument as "Meanings ain't in the head."[29] In my opinion, the lesson to learn from Putnam's argument is not that cognitive semantics are impossible but that their proponents have forgotten about the *social structure* of language.

In this book, a semantic theory based on *meetings of minds* will be presented. According to this view, the meanings of expressions do not reside in the world or (solely) in the image schemas of individual users but *emerge from the communicative interactions of language users*. Consequently, meanings are in the *heads* of the users (Gärdenfors, 1993; Warglien & Gärdenfors, 2013).[30]

The central claim is that the social meanings of the expressions of a language are indeed determined from their individual meanings, that is, the meanings that the expressions have for the individuals.[31] It is important to note that I am not assuming that individuals have the same image schemas or representations of the world.

1.5.2 Slow and Fast Meetings of Minds

The fundamental role of human communication is, indeed, to affect the state of mind of others, bringing about cognitive changes (van Benthem, 2008). Recall that to communicate means "to make common." A meeting of minds means that the representations in the minds of the communicators become sufficiently compatible to satisfy the goals that prompted the communication.

There are two basic types of meetings of minds: one *slow* and one *faster*. The slow one concerns how a community adjusts its uses of words, gestures, and so on, so that they obtain relatively fixed meanings within the community that are largely independent of any particular communicative context. In chapter 5, I show how this process can be modeled in terms of *equilibriums* in language games, together with the *geometry* and *topology* of concepts.

The fast type of meetings of minds concerns expressions that obtain their meaning during a communicative interaction. It can sometimes be achieved without the aid of language. As an example, consider *declarative pointing* (Bates, 1976; Brinck, 2004a; Gärdenfors & Warglien, 2013). This consists of one individual pointing to an object or location and, at the same time, checking that the other individual (the recipient) focuses her attention on the same object or location. The recipient, in turn, must check that the pointer notices the recipient attending to the right entity (fig. 1.5). This "attending to each others' attention" is known as *joint attention* (Tomasello, 1999).

The fast process concerns the development of a shared world—*common ground*—during a dialogue or a similar exchange of communicative acts.[32] Several researchers (Stalnaker, 1978; H. Clark, 1992) have emphasized that human communication to a large extent relies on the interlocutors' sharing a common ground. Clark (1992, pp. 4–5) says that this is characterized by the following properties:[33]

(i) The participants in a conversation work together against a background of shared information.
(ii) As the discourse proceeds, the participants accumulate shared information by adding to it with each utterance.

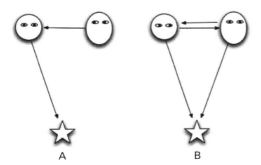

A   B

Figure 1.5
Two steps in achieving joint attention.

(iii) Speakers design their utterances so that their addressees can readily identify what is to be added to that common ground.

In this process, the meanings of the community are taken as a starting point, but to this are added the meanings of pronouns, definite descriptions, and so on, that to some extent depend on the communicative environment. The fast process rapidly accumulates new meanings that the exchange can build on in later stages.

Clark's notion introduces an ambiguity, since it contains both common background knowledge that is comparatively stable between speakers and the alignment of information that is created during a conversation (cf. Pickering & Garrod, 2004). These two forms of common ground should be kept separate.

When one speaks of the semantics of a language, normally only the slow process is treated. Aspects of communication that belong to the fast process are seen as belonging to pragmatics. However, I want to argue that large similarities exist between the slow and the fast forms of meaning construction. There are words that change their meanings slowly, and there are words that change their meanings rapidly in the course of a dialogue. For example, in spite of all scientific advances, the meaning of "stone" has presumably not changed much since the Stone Age.[34] On the other hand, there are several components of language, for example, pronouns, that only receive meaning during the fast process. For example, the meaning of the word "this" changes almost every time it is used. Consequently I do not draw any sharp border between semantics and pragmatics (cf. Langacker, 2008, pp. 40–41).

I present the theory of semantics as meetings of minds in greater detail in chapter 5. The theory is sociocognitive in the sense that it takes the communicative acts as part of the process of building meanings, while at the same time using conceptual spaces to model the inner worlds of the communicators. One goal of the first part of the book is to present a theory where the mechanisms of forming and updating the common ground can be modeled. I will also show that the model satisfies all the six criteria presented earlier.

This chapter has presented the setting for the rest of the book. In chapter 2, I introduce my primary tool for modeling semantic concepts and processes, that is, the theory of conceptual spaces.

2 Conceptual Spaces

2.1 The Geometry of Meaning

Our words express our concepts. Hence a theory of semantics should be founded on a theory of concepts. Croft makes the connection as follows:[1]

The categories defined by constructions in human languages may vary from one language to the next, but they are mapped onto a common conceptual space, which represents a common cognitive heritage, indeed the geography of the human mind (Croft, 2003, p. 139) . . . which can be read in the facts of the world's languages in a way that the most advanced brain scanning techniques cannot ever offer us. (Croft, 2001, p. 364)

This book, however, focuses not on the geography of the mind but on its *geometry*. My modeling tool is the theory of conceptual spaces that I have previously developed (Gärdenfors, 2000). A central idea is that the meanings that we use in communication can be described as organized in abstract spatial structures that are expressed in terms of *dimensions, distances, regions*, and other geometric notions. In addition, I also use some notions from *vector algebra*.

I begin by summarizing the theory and some of its developments. Conceptual spaces are constructed out of *quality dimensions*. Examples are pitch, temperature, weight, size, and force. The primary role of the dimensions is to represent various "qualities" of objects in different domains. Some dimensions come in bundles—what I call *domains*—for example, space (dimensions of height, width, and depth); color (hue, saturation, and brightness); taste (salt, bitter, sweet, and sour, and maybe a fifth dimension); emotion (arousal and value; see sec. 3.2.1); and shape (dimensions not well known; see sec. 6.3). These dimensions are closely connected to what is involved in perceiving and acting (see, e.g., Schiffman, 1982). However, there also exist quality dimensions that are of an abstract, nonsensory character.

The notion of a dimension should be understood literally. It is assumed that each dimension is endowed with a *topological* or *geometric* structure. As a first example, take the dimension of time. Time is a one-dimensional structure that is isomorphic to the line of real numbers. If now is seen as the zero point on the line, the future corresponds to the positive real line, and the past to the negative line. In contrast, the dimension of weight is isomorphic to the positive real line; there are no negative weights. As I show in chapter 7, the differences in dimensional structures show up in the semantics of the adjectives that are used to describe the dimensions.

Topology and geometry allow us to talk about *nearness* and *distance* in a space. The interpretation is that if point x is nearer to point y than to point z, then x *is more similar* to y than to z. Similarity relations will be important in the representation of concepts.[2] If it is assumed that a domain has a metric, one can put numbers on distances in the space. Note, however, that for many domains, it is not meaningful to measure distances.

It is important to introduce a distinction between a *cognitive* and a *scientific* interpretation of quality dimensions. The cognitive interpretation concerns the structure of human perception (and that of other animals) and their inner worlds. The scientific interpretation, on the other hand, deals with how different dimensions are presented within a scientific theory. For example, in optics the hue of a color is determined by a linear wavelength dimension, while the psychological representation is a circular dimension (see fig. 2.1).[3] When the dimensions are seen as cognitive entities, their structure should be determined not by scientific theories that attempt to give a "realistic" description of the world but by *psychophysical* measurements that determine how concepts are represented in our minds.

Since the notion of a domain is central to my analysis, it should be given a more precise meaning. To do this, I rely on the notions of separable and integral dimensions from cognitive psychology (Garner, 1974; Maddox, 1992; Melara, 1992; Kemler Nelson, 1993). Certain quality dimensions are *integral* in the sense that one cannot assign an object a value on one dimension without giving it a value on the other. For example, color perception consists of integral dimensions: an object cannot be given a hue without also giving it a saturation value. Or the pitch of a sound always goes along with a specific loudness. Dimensions that are not integral are said to be *separable*, for example, the size and hue dimensions.[4]

Using this distinction, I define a *domain* as a set of integral dimensions that are separable from all other dimensions. (Many domains, such as temperature or weight, consist of only one dimension.) The most fundamental semantic reason for decomposing a cognitive structure into domains is that the properties assigned to a domain can be described *independently* of other properties of an object. For example, an object can be assigned the weight one kilo independently of its temperature or color. A main theme of this book is that the division of cognitive representations into domains is strongly reflected in semantics. I discuss the notion of a domain further in section 2.5.

A psychologically interesting example of a domain is *color*. As already mentioned, our cognitive representation of color can be described by three dimensions. The first dimension is *hue*, which is represented by the familiar *color circle*. The topological structure of this dimension is thus different from the quality dimensions representing time or weight, which are iso-morphic to the real line.

The second psychological dimension of color is *saturation*, which ranges from gray (zero color intensity) to increasingly greater intensities. This dimension can be mapped onto an interval of the real line. The third dimension is *brightness*, which varies from white to black and is thus a linear dimension with endpoints. Together these three dimensions, one with circular structure and two with linear, constitute the color domain that is a subspace of our perceptual conceptual space. The domain is often illustrated by the so-called *color spindle* (fig. 2.1). Brightness is shown on the vertical axis. Saturation is represented as the distance from the center of the spindle. Hue, finally, is represented by the positions along the perimeter of the central circle. The reason that the color space forms a spindle and not a cylinder is that it is more difficult to discriminate colors close

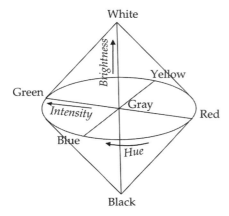

Figure 2.1
The color spindle.

to white or black. For example, a very light green is much more similar to a very light blue than green is to blue.

In connection with the dimensions of the color space, I would like to comment on their ontological status. I view them as *theoretical constructs* that are instrumental in systematizing and explaining the color similarity judgments.[5] Other accounts of the color space use other dimensions, and the question of which model provides the best description of the data is still debated. I should also emphasize that the dimensions are seen as cognitive constructs, and they should not be mapped onto wavelengths or any other physical properties. The dimensions of the other domains that I discuss here have the same status. Whether the dimensions have a correspondence in the brain in the sense that they correspond to some neural structure or neural mechanism is a different question that I will not attempt to answer.[6]

It is impossible to provide a complete list of the domains involved in the conceptual spaces of human cognition.[7] Some of the domains develop early in childhood. Other dimensions are learned later in life. Learning new concepts often involves expanding one's conceptual space with new domains, as we shall see in chapter 3.[8] As Waismann (1968, p. 121) noted, most concepts have an essential incompleteness in the sense that one may describe a number of properties of the concept, but the description will never be complete. In chapter 3, I present an analysis of the main domains that are required for the development of communication during a child's first years.

2.2   Properties versus Concepts

Asking a psychologist, philosopher, or a linguist what a concept is is much like asking a physicist what mass is. An answer cannot be given in isolation.
—Ray Jackendoff

The general description of conceptual spaces will now be used to introduce a distinction between *properties* and *concepts*. I use the notion of a property to denote information related to a single domain. The following thesis was proposed in Gärdenfors (1990, 2000), where the geometric characteristics of the quality dimensions are used to introduce a spatial structure to properties:

Thesis about properties: A *property* is a convex region in some domain.

That a region R is convex means that for any two points x and y in R, all points *between* x and y are also in R.[9] The motivation for the thesis is that if some objects located at x and y in relation to some domain are both examples of a property, then any object that is located between x and y with respect to the same domain will also be an example of the property. Although not all domains in a conceptual space may have a metric, I assume that the notion of betweenness is defined for all domains.[10] This will make it possible to apply the thesis about properties generally.

Properties, as characterized by the thesis, form a special case of *concepts*. This distinction is defined by saying that a property is based on a single domain, while a concept is based on one *or more* domains (Gärdenfors, 2000). The distinction has been obliterated in both linguistic and philosophical accounts. For example, properties, concepts, and relational concepts are all represented by predicates in first-order logic and in λ-calculus.

In my earlier work (Gärdenfors, 2000), I only considered concepts in the form of *object categories*.[11] In this book, the notion of a concept will be extended to action and event categories. Furthermore, in section 6.7, I discuss *relational concepts* (Gentner & Kurtz, 2005), such as *passenger* or *team leader*, which can only be defined in relation to other concepts. Nevertheless object categories are still central examples of concepts.

Concepts are not just bundles of properties. The representation of an object category that I present in chapter 6 also includes an account of the *correlations* between the regions from different domains that are associated with the category. For example, the concept *apple* involves a very strong (positive) correlation between the sweetness in the taste domain and the sugar content in the nutrition domain and a weaker correlation between the color red and a sweet taste.

An example of a concept that is represented in several domains is *bird*.[12] A bird has a shape, a size, colors, a weight, a sound, an ecological habitat, and so on. Langacker (1987, p. 152) calls the set of domains that are used to characterize a concept the *domain matrix*. The set of domains for a concept may not be closed but can be expanded as more knowledge is accrued.

Concepts have multiple cognitive functions: they help in organizing and categorizing our perceptions, they are involved in reasoning, and they form the basis for the meanings of words. In this book, I focus on the role of concepts in semantics. I see it as a strength that I show how my model of concepts also connects to perception and reasoning.[13]

2.3   The Convexity Requirement

Geometry . . . may be called the general science of classification.
—Alfred North Whitehead

That properties are convex may seem a strong assumption, but convexity is a remarkably regular property of many perceptually grounded domains: for example, color, taste, vowels, musical scales (Gärdenfors, 2000; Jäger, 2008; Honingh & Bod, 2011). As I argue in chapter 8, it applies to our categorization of actions as well.

The main argument used in Gärdenfors (2000) to support the convexity requirement is that it facilitates the *learnability* of categories. An empirical test of this position was performed by Lee and Portier (2007) using multi-dimensional data from 170 pieces of fruit of seventeen different kinds. The fruit data were represented by five domains: color, shape, size, weight, and squash measurements.[14] The researchers compared learning in convex regions in conceptual spaces to other methods from machine learning using multidimensional feature spaces. Their results showed that when convex regions were used, the average learning accuracy was 97 percent, while for the other methods the average accuracy was 56 percent. This result confirms that the convexity of regions provides a robust and efficient framework for concept learning in artificial systems.

In this book, I will show that the convexity of concepts is also crucial for ensuring the *effectiveness of communication*. As we shall see, the learn-ability and effectiveness of communication clearly interact in complementary ways in the acquisition of concepts. Strong support for this idea has been provided by Jäger (2008, p. 552). He argues that "languages where meanings are convex regions (of a special kind) are . . . optimally adapted to communication. The preference for convex meanings can thus be seen as the result of some process of (cultural) evolution." I return to his argument in section 7.2 when I discuss color properties.

There are interesting comparisons to make between analyzing properties as convex regions and the prototype theory developed by Rosch and her collaborators (see, e.g., Rosch, 1975, 1978; Mervis & Rosch, 1981; Lakoff, 1987). When properties are defined as convex regions in a domain, proto-type effects are to be expected. Given a convex region, one can describe positions in that region as being more or less central.[15]

Conversely, if prototype theory is adopted, then the representation of properties and concepts as convex regions is to be expected. Assume that some metric quality dimensions of a conceptual space are given—for example, the dimensions of the color domain—and that the goal is to decompose it into a number of regions, in this case, color properties. If one starts from a set of prototypes $p_1, \ldots, p_n$, these should be the central points in the properties they represent. The information about prototypes can then be used to generate convex regions by stipulating that any point

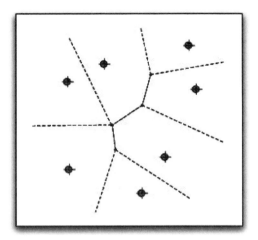

Figure 2.2
A Voronoi tessellation of the plane into convex sets.

p within the space belongs to the same property as the *closest* prototype p_i. (Note that for us to talk about a "closest" point, the spaces must have a metric.) This rule will generate a *Voronoi tessellation*. Figure 2.2 depicts a Voronoi tessellation. The particular tessellation is two-dimensional, but Voronoi tessellations can be extended to any number of dimensions.

A central semantic hypothesis is that the most typical meaning of a word or linguistic expression is the prototype at the center of the convex region assigned to the word.[16] However, note that concepts such as *tall* and *warm* do not have prototypes. They correspond to open-ended regions of a dimension, where no point can be identified as the most typical.

It is easy to prove that a Voronoi tessellation always results in a decomposition of the space into convex regions (Okabe, Boots, & Sugihara, 1992). The tessellation provides a geometric answer for how a similarity measure together with a set of prototypes can determine a set of categories. The partitioning results in a *discretization* of the space. Having a space partitioned into a finite number of regions means that a finite number of *words* can be used to refer to the regions. However, as I discuss in section 2.8, psychological metrics are imprecise and often context dependent. As a consequence, the borderlines will not be exactly determined.

The tessellation mechanism provides important clues to the *cognitive economy* of concept learning. If the categorization of each point in a space had to be memorized, this would put absurd demands on human memory. However, if the partitioning of a space into categories is based on a Voronoi tessellation, only the relative positions of the prototypes need to be remembered. Once you recall the positions of the prototypes, the rest of a categorization can be computed by using the metric of the space. In brief, the tessellation mechanism discretizes the world and in this way it relieves memory.[17]

A Voronoi tessellation depends on the *level of specificity* of the class of concepts represented by the tessellation. Rosch (1975, 1978) distinguishes between superordinate, basic, and subordinate levels. Each level will generate coarser of finer tessellations. A natural constraint on a conceptual hierarchy is that that the region associated with a higher-level category includes all the regions of concept meanings on lower levels.[18]

2.4 Product Spaces

Descartes invented the coordinate system as a way of representing the connections between two variables x and y. He thereby introduced a powerful visualization of a *product space*. From two independent dimensions X and Y, a space $X \times Y$ that consists of all pairs $<x,y>$ of points from X and Y can be created. These pairs can be depicted in a two-dimensional Cartesian coordinate system. This construction can be generalized to spaces (domains) X and Y with several dimensions.

Whenever dimensions from different domains are combined or compared, then the product space of the domains becomes relevant. For example, *relations* can be analyzed as product spaces (see Gärdenfors, 2000, sec. 3.10.1). To illustrate, consider the semantics of *comparative adjectives*. A comparative adjective like "warmer" refers to the temperatures of two objects. The temperature dimension can be represented by the dimension R of real numbers. If $t(x)$ denotes the temperature of an object x, one can then consider the product space of all pairs of temperature values $<t(x),t(y)>$, that is, R^2. In this space, the relation "x is warmer than y" can be identified with the region of all pairs such that $t(x) > t(y)$. This region turns out to be a *convex* subset of R^2 (all points below the diagonal in the product space). Thus the thesis about the convexity of properties can, in a natural way, be extended to a thesis about relations. It is easy to see that this kind of analysis applies to all comparative relations where the generating dimension is isomorphic to the real numbers.

It should be noted, however, that convex regions in product spaces are not just the products of the convex regions of the underlying dimensions. To give an example from Adams and Raubal (2009), the concepts *hill* and *mountain* are to a large extent dependent on the values for the width and

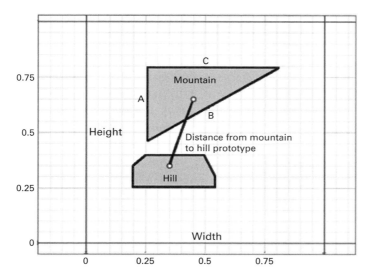

Figure 2.3
Representing *hill* and *mountain* in the product space of the width and height dimensions. © 2009 IEEE. Reprinted, with permission, from B. Adams and M. Raubal. (2009). The Conceptual Space Markup Language (CSML): Towards the Cognitive Semantic Web. In *Third IEEE International Conference on Semantic Computing* (pp. 253–260). Berkeley, CA: IEEE Computer Society.

height dimension of the geographic objects (fig. 2.3). If a formation is very high, its width will not matter much; it will still be a mountain. However, a lower and very wide formation might not be called a mountain. Thus the region in the product space that represents *mountain* has more or less a triangular shape. (If the region had been a product of convex regions of the width and height dimensions, it would have had a rectangular shape.)

Several examples of constructions of product spaces can be identified in child development. The most famous is perhaps Piaget's (1972) conservation task. Young children tend to identify the volume of a liquid with the height of its container, for example, a glass. Only when they are older are they able to represent volume as a product of both the height and the width of the container.[19]

In product spaces, *correlations* between dimensions can be detected. If the correlation between two dimensions is strong, then it is sometimes possible to identify a new (higher-level) dimension that can form the basis for new meanings. In this way, new concepts can be created as regions in

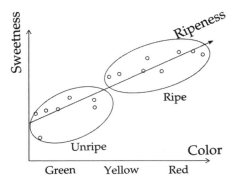

Figure 2.4
Ripeness as a correlational dimension.

product spaces. For example, a strong correlation exists between the sweetness values and the colors of fruits. The correlation will show up as a more or less linear structure in the product of the taste and color domains. This line can then be employed as a new semantic dimension of "ripeness," where regions for *ripe* and *unripe* can be separated out (fig. 2.4).

I use product spaces as representational tools at a number of places in this book. In particular, in chapter 4 I consider several examples in relation to different kinds of pointing, and in chapter 13 in relation to compositionality.

2.5   Domains in Cognitive Linguistics

As I have already presented for the case of colors, dimensions do not come alone but are often organized in groups—*domains*—belonging to the same kind of meanings.[20] This book presents a model of such domain-oriented language learning.[21] In the following chapter, I illustrate the model with some of the semantic domains that a child acquires during her first formative years.

Also within cognitive linguistics in general, the notion of a domain is central. A number of articles analyze the meaning of the notion (e.g., Croft, 2003; Clausner & Croft, 1999; Croft & Cruse, 2004; Engberg Pedersen, 1995; Evans & Green, 2006), but they all fall back, more or less, on chapter 4 of Langacker (1987).

Following Langacker (1987), the literature shows a strong tendency to interpret the notion of domain in an all-encompassing way. Langacker (2008, p. 44) writes that "the term is broadly interpreted as indicating any kind of conception or realm of experience." Another example is Clausner and Croft (1999, p. 1), who write: "Although the basic constructs of 'concept,' 'domain,' 'construal,' and 'category structure' go by different names, they are essentially the same among researchers in cognitive linguistics."[22] One reason for this sameness is that the early cognitive semantics made a key distinction between "figure" and "ground" that was borrowed from Gestalt psychology. A basic idea is then that the semantic structure for a word (or a construction) consists of both the concept (figure) and its presupposed domain structure (ground).[23] In my opinion, the figure-ground distinction is not appropriate in all situations, and using it as a universal principle leads to unnecessary confusion. My aim in this section is to clear up some of this conceptual complexity.

2.5.1 Basic and Abstract Domains

I first critically discuss some attempts to specify what is meant by a domain. I argue that Langacker's concept of domain conflates several components that are invoked in the analysis of lexical meanings. In particular, I believe that his distinction between locational and configurational domains is misleading. His configurational domains are better seen as *meronomic* information concerning the relation between parts and whole, rather than something pertaining to domains.

Langacker (1987) wants to show that meaning is based on conceptualization: his motivation for introducing the notion of domain is part of this program. He says that domains "are necessarily cognitive entities: mental experiences, representational spaces, concepts or conceptual complexes" (1987, p. 147). He describes a domain as a "context for the characterization of a semantic unit." This description is in accordance with viewing a domain as a ground in the Gestalt sense.

The first distinction Langacker discusses concerning domains is that of *basic* versus *abstract* domains. He notes that the concept *knuckle* "presupposes" *finger*, which, in turn, depends on *hand*, *arm*, *body*, and finally *space* (1987, pp. 147–148). However, *space* cannot be defined relative to some other more fundamental concept. This is what characterizes a *basic domain*.[24] Langacker (1986a, p. 5) summarizes:

It is . . . necessary to posit a number of "basic domains," that is, cognitively irreducible representational spaces or fields of conceptual potential. Among these basic domains are the experience of time and our capacity for dealing with two- and three-dimensional spatial configurations. There are basic domains associated with various senses: color space (an array of possible color sensations), coordinated with the extension of the visual field; the pitch scale; a range of possible temperature sensations (coordinated with positions on the body); and so on. Emotive domains must also be assumed. It is possible that certain linguistic predications are characterized solely in relation to one or more basic domains, for example time for (BEFORE), color space for (RED), or time and the pitch scale for (BEEP). However, most expressions pertain to higher levels of conceptual organization and presuppose nonbasic domains for their semantic characterization.

Evans and Green (2006, p. 234) add that basic domains are directly tied to preconceptual embodied experience and provide "a 'range of conceptual potential' in terms of which other concepts and domains can be understood." They are fundamental for understanding the world. As I argue in the following chapter, many of the domains that are learned early in development are consequently basic domains in this sense.

An *abstract domain* is then defined as "any nonbasic domain, i.e., any concept or conceptual complex that functions as a domain for the definition of a higher-order concept" (Langacker, 1987, p. 150). In a footnote to the definition, he adds: "An abstract domain is essentially equivalent to what Lakoff . . . terms an . . . idealized cognitive model . . . and what others have variously called a frame, scene, schema or even script."

The problem with this general characterization, which is repeated in most later discussions, is that it is too inclusive: it offers no criterion for what is *not* a domain. Taylor (1989, p. 84) writes that "in principle, any conceptualization or knowledge configuration, no matter how simple or complex, can serve as the cognitive domain for the characterization of meanings." The all-embracing description of domains makes the concept too vague and rather useless. In my opinion, it is necessary to separate meaning relations that are based on *similarity judgments* from other types of relations. I suggest a narrower characterization of domain based on dimensionality and argue that many other aspects of meaning that have been sorted under the notion instead concern part-whole and other meronomic relations.

2.5.2 Locational and Configurational Domains

One of the central, but also more problematic, distinctions in Langacker's (1987) theory is between *locational* and *configurational* domains.[25] Langacker's description is ambiguous. First he writes that a *concept* "can be characterized as a location or a configuration in a domain (or in each of a set of domains)." A color is locational, since it can be identified with a *region* of color space, while a *circle* is configurational, since it can be described as a configuration of points in the domain of two-dimensional space (Langacker, 1987, p. 149).

Then he extends the terminology to domains. "A predicate specifies a location or a configuration in a domain (or in each domain of a complex matrix). Accordingly we can speak of a domain being either locational or configurational" (Langacker, 1987, p. 152). Langacker discusses possible criteria for determining whether a domain is locational or configurational. He settles on the following: "What makes a domain configurational . . . is our capacity to accommodate a number of distinct values as part of a single gestalt" (1987, p. 153). Using this criterion, he concludes that space, time, and pitch are configurational domains, while color is locational.

Clausner and Croft (1999, sec. 2.2) criticize this distinction, showing that the domains Langacker calls configurational can support both locational and configurational concepts:[26] space supports locational *here* and configurational *circle*, time supports locational *now* and configurational *week*, and pitch supports locational *middle C* and configurational *minor chord*.

Clausner and Croft (1999, p. 13) take a step in the right direction when they recognize that "locationality vs. configurationality is a property of concepts, not domains." I want to go further: I shall show that by considering "higher-level" domains, concepts described as configurational in one domain can be seen as locational in another. This means that once a hierarchy of domains is introduced, the distinction loses much of its relevance.

2.5.3 Dimensionality

As a third theme in his presentation of domains, Langacker (1987, pp. 150–152) observes that many domains—basic and abstract—are *dimensional*, but he does not formulate this property as a criterion of a domain. The examples Langacker and others give of domains can be grouped into two categories: (i) *dimensional* and (ii) *meronomic* (that is, relating to part-whole relations). Langacker's first example—*finger* as the domain for *knuckle*, *hand* as the domain for *finger*, and so on—is a clear case of meronomic structure. Similarly, when he claims that no *hypotenuse* exists without a *right-angled triangle* (Langacker, 1987, p. 183), this is a statement about a meronomic relation. I interpret Langacker's (1987) proposal to distinguish between locational and configurational domains as his attempt to characterize the two types of domains. Thus formulated, the distinction is misleading.

Clausner and Croft (1999, p. 6) go so far as to say that "the concept-domain semantic relationship is essentially a part-whole (i.e., meronomic) relationship." This may only be true in a different sense for dimensional domains: concepts correspond to *regions* of dimensional domains. This is not normally what is meant by "a part-whole relationship."

Another distinction is more subtle. When Langacker writes that one concept *presupposes* another, this can mean two different things. Although a finger presupposes a hand, the meronomic structure makes it possible to replace one finger with another, with the result still being a hand. Or the engine of a car can be replaced with another, and the object still remains a car. In a dimensional domain like color—in contrast—*orange* might be said to presuppose the color domain; but orange cannot be replaced by another color while keeping the content of the domain the same. Therefore the color domain does not have a meronomic structure. It forms an integrated whole where the *relations* between the elements are central for determining the content of the domain: orange must be a color *between* yellow and red, or else it is no longer the color domain. In economic terms, the parts in a meronomic structure are *fungible*, while regions of dimensional domains are not.

Meronomic relations do not occur only in the shape domain. They can be found in many other domains: a chord consists of three tones or more; a family, of parent(s) and child(ren); a limerick, of five lines; and so on.

When Langacker (1987, p. 183) talks about the concept *arc* being "profiled" against the "base" *circle*, he also uses the term "domain" for the base. He emphasizes the distinction between profile and base in many situations. However, all his profile-base examples involve meronomic relations. Croft (2003, p. 166) explicitly relies on the profile-base distinction in defining a domain as "a semantic structure that functions as the base for at least one concept profile." I propose to take the other direction and define domains in terms of dimensions.

The dimensional domains have different *semantic roles* than do the meronomic structures. The examples given for meronomic structures are of concepts expressed by nouns; thus they refer to objects and parts of objects. Characteristically, objects have multiple properties, while a dimensional domain only represents a single property (see chap. 6). In cognitive linguistics, metaphors have been characterized as a mapping across domains and metonymies as involving relations within a domain. However, the broad use of the notion of a domain has made it difficult to apply this distinction. An additional benefit of the distinction between dimensional domains and meronomic relations is that it generates a natural account of the difference between metaphors and metonymies (see sec. 2.6).

Clausner and Croft (1999, p. 5) note that "Langacker's notion of a domain appears to differ in some respects from the term *domain* used by most psychologists dealing with concepts." I now turn to a brief description of the notion of a domain as used in cognitive psychology. I aim to show that the psychological notion of domain is a suitable tool for the aims of cognitive linguistics. Thus I propose to unify the two areas.

Even if we do not know much about the geometric structures of many of the domains, it is quite obvious that there is some such nontrivial structure. Psychology has a long tradition of analyzing the dimensionality of perceptual structures. Shepard (1987) argues that as soon as perception relating to a particular domain can be graded by similarity, mathematical techniques—such as multidimensional scaling (Kruskal, 1964) or principal component analysis—can be used to extract a low-dimensional space representing the similarity judgments. The more similar two objects are judged to be, the closer the points representing the objects will be located within the space. A large number of perceptual spaces have been identified in this way (see, e.g., Shepard, 1987, p. 1318).[27]

I submit that everything Langacker calls a basic domain can be given a dimensional description. The claim must, however, be qualified: some dimensions are binary (e.g., gender), some are ordered structures (e.g., kinship relations), and some have a full metric structure.

As a potential counterexample, Langacker (1987, p. 151) says that emotions cannot be characterized "in terms of a small number of essentially linear dimensions."[28] However, one finds several dimensional analyses of emotions in the psychological literature. Most of them (e.g., Osgood, Suci, & Tannenbaum, 1957; Russell, 1980) contain two basic dimensions: a *value* dimension, on a scale from positive to negative aspects of emotions; and an *arousal* dimension, on a scale from calm to excited emotional states.[29] Although there is no one final theory of the dimensional structure of emotions, I do not view the emotion domain as a counterexample to my thesis. I discuss the emotion domain further in section 3.2.1.

To summarize, results from cognitive psychology support the thesis that basic domains are dimensional. I next want to show that what have been called configurational domains can likewise be given a dimensional analysis as "higher-level" domains.

2.5.4   Configurations Are Higher-Level Domains

Certain properties can be described as more general patterns arising from relations between points in a conceptual space.[30] I have already discussed the binary relations such as *warmer* that are expressed by comparative adjectives. In an ordinary two-dimensional space (R^2) with standard Euclidean metric, one can, of course, define regions as subsets of the space. One

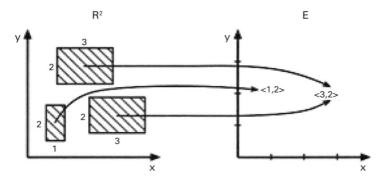

Figure 2.5
The configurational space E of rectangles.

can also introduce a new space of "shapes" that are patterns of points in the space. Consider the concept *rectangle*, a typical configurational concept, according to Langacker. It can be defined using the set of all quadruples $<a,b,c,d>$ of points in R^2 satisfying the condition that the lines ab, bc, cd, and ad form a convex polygon, with $a_x - b_x = c_x - d_x$ and $a_y - c_y = b_y - d_y$ (i.e., the sides are pairwise equally long), and $|ad| = |bc|$, where $|ad|$ and $|bc|$ denote the length of the diagonals (i.e., the diagonals are equally long). One can partition this set of quadruples into equivalence classes by saying that two rectangles $<a,b,c,d>$ and $<e,f,g,h>$ are identical if $|ab| = |ef|$ and $|ac| = |eg|$. In this way, one identifies a rectangle by the length of its sides, independent of its position in R^2.[31] Now, let E be the set of all such equivalence classes.[32] This amounts to mapping the rectangle $<a,b,c,d>$ to the point $(|ab|,|ac|)$ in R^2 (fig. 2.5). The space E of rectangles is therefore isomorphic to $(R^+)^2$, with the width and the length of the rectangles as the generating dimensions.

Within E, one can now identify regions that correspond to certain properties. For example, a *square* is any point $<|ab|,|cd|>$ in E such that $|ab| = |ac|$. It is easy to show that this region is a convex subset of E, since that subset corresponds in E to the line $x = y$, and a line is a convex set. On the right side of the line (i.e., where $y > x$), one finds the set of all rectangles taller than they are wide: for example, the rectangle $<1,2>$ in figure 2.5. This set is also convex (fig. 2.6).

This analysis of the concept of rectangle is an elementary example of how configurations (patterns) can be defined as higher-order correlations in an underlying space. The question remains, of course, to what extent this account of configurational concepts fits with our perceptual representations, about which we know lamentably little.

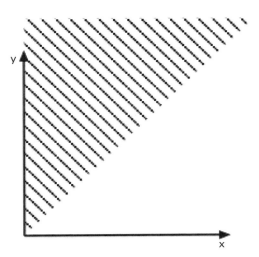

Figure 2.6
The space of vertically extended rectangles.

In this way, an object can be represented as a pattern (vector) in a high-dimensional space, and the shape space is *supervenient* on the spatial and angular dimensions, just as rectangles are supervenient on the length and width dimensions. The supervenience relation introduces a partial hierarchy of domains that should be considered in any lexical analysis of concepts.

The argument concerning the concept of rectangle can, in principle, be generalized to other shape concepts. If we adopt something like Marr and Nishihara's (1978) analysis, we can begin to see how such a space would appear. In chapter 6, I outline their approach to the shape domain. Furthermore, in chapter 8, I show how the force domain forms the basis for the configurational domain of actions.

This mechanism can be generalized to other allegedly configurational concepts. Thus I submit that even these concepts can be related to dimensional domains. In brief, the distinction between locational and configurational domains is not needed but can be expressed with the aid of basic and higher-level dimensional domains.

Consider an example from another type of domain: Clausner and Croft (1999, p. 9) argue that *intervals* and *chords*—on the pitch dimension—are configurational concepts. A similar analysis to that of rectangles can easily be developed for *interval* and *chord*. However, I will not pursue the details here.

	0	1	2	3
0	I	mother	grandmother	great-grandmother
1	daughter	sister	aunt	(grand)aunt
2	granddaughter	niece	cousin	cousin
3	great-granddaughter	grandniece	cousin	cousin

Figure 2.7
Diagram based on Zwarts's two-dimensional product space of kinship relations (only one gender is represented in this figure).

Another example is the domain of *kinship relations*, generated by combining links of the type "*x* is a child of *y*," "*x* is a parent of *y*," "*x* is married to *y*," "*x* is a sister or brother of *y*," to generate a genealogical tree structure. The partitioning of that tree generated by kinship words varies widely between languages: for example, many cultures make no distinction between *mother* and (maternal) *aunt*.

A genealogical tree is not the only way to represent kinship relations. Zwarts (2010a) proposes a two-dimensional product space of kinship relations. One dimension is the number of steps backward through the genealogy, represented by the horizontal dimension in figure 2.7; the other is the number of steps forward, represented by the vertical dimension.

Note that—in accordance with the property thesis—this representation makes *cousin* correspond to a convex region, unlike what happens when the concept is represented in a genealogical tree. In this way, one sees that, depending on which representation is used, concepts become differently related.

The upshot is that many higher-level domains can be given a dimensional analysis. An even stronger claim would be that this holds for all domains, both basic and higher level. However, such a claim can only be taken as programmatic, since we do not have sufficient empirical evidence, so far, to evaluate it. Nevertheless traditional semantic notions such as antonymy and hyponymy provide material for sorting concepts into domains: if two concepts *A* and *B* are antonyms of each other (e.g., *dry* and *wet*), or if *A* is a hyponym of *B*, (e.g., *terrier* and *dog*), then *A* and *B* belong to the same domain. Even if it turns out that separable domains are difficult to identify, the distinction between dimensional domains and meronomic structures is important for lexical semantics, as I argue in chapters 6 and 7.

2.6 Metaphors and Metonymies

In contrast to many other semantic theories, cognitive linguistics has brought forward the importance of metaphors and metonymies as strong tools for expanding the meanings of words. In chapter 11, I argue that these mechanisms are indeed the most important forms of meaning transformations.

Linguists and philosophers disagree, however, on how to characterize metaphors and metonymies. As an indication, Broström (1994, pp. 27–28) provides the following example of how the notion of domain as it has been used in cognitive linguistics leads to problems concerning how metaphors are to be identified:

If when looking at a dog, I think of his *face*, is that the result of metaphorical categorization? When looking at a caterpillar? Or, to recast the question in domain terms, does the concept of a face belong to the domain of the human body, to the more general domain of animate bodies or to a domain of intermediate scope, say mammalian bodies?

She analyzes the problem as follows (see also Jackendoff & Aaron, 1991):

Now, even if the concept of a domain may be a useful approximation, as is the term "metaphor" itself, it does not provide us with an explanation of the difference between litera [literal meaning] and metaphor. The reason why is that it is not possible to individuate domains, to tell one domain from another, independently of establishing which expressions are literal and which metaphorical. Domain boundaries are not clearer than the boundary between litera and metaphor. (Broström, 1994, p. 26)

This diagnosis supports my claim that the use of the notion of domain in cognitive linguistics is too lax. Given Broström's analysis, I propose that "face" is better analyzed in meronomic terms. Given this distinction, the boundary between literal meaning and metaphor will become sharper.

2.6.1 Characterizing Metaphors and Metonymies

On the basis of conceptual spaces, it is natural to say that a metaphor expresses an identity in structure between different domains. A word representing a particular pattern in one domain can be used as a metaphor to express the same pattern in another domain. In his *invariance principle*, Lakoff (1993, p. 215) formulates a closely related position: "Metaphorical mappings preserve the cognitive topology (that is, the image-schema structure) of the source domain, in a way consistent with the inherent structure of the target domain." In this way one can account for how a metaphor transfers information from one conceptual domain to another; what is transferred is a pattern rather than domain-specific information. A metaphor can thus be used to identify a structure in a domain that would not have been discovered otherwise. This is how metaphors create new knowledge. By exploiting geometric and topological invariances, the theory of conceptual spaces provides a principled way of describing the mappings.[33]

The role of metonymy is primarily referential. By picking out one specific aspect of an entity, metonymy focuses attention on something salient to the situation, thereby helping one's understanding of it. Classical types of metonymy include *pars pro toto*, where the focus is on a part of an object instead of the whole; and *totum pro parte*, where the focus is on something that contains an object as one part. The relevant parts and wholes need not just be spatial but can be temporal as well. More abstract meronomic relations of a functional or causal nature also generate metonymies: for example, in the sentence "Proust is tough to read," the author stands for the book he wrote; while in "Napoleon attacked Russia," the highlight is Napoleon's function as leader of the army.[34] (For an extensive list of different types of metonymies, see Peirsman & Geeraerts, 2006, sec. 2.)

Metonymic concepts make it possible to conceptualize something by its relation to something else to which it is connected. In contrast to metaphor, metonymy is based not on the similarity between two domains but on meronomic (and other) relations *within* the same domain. Here I agree with Lakoff and Turner (1989, p. 103), who argue that—unlike metaphor—metonymy "involves only one conceptual domain. A metonymic mapping occurs within a single domain, not across domains" (see also Peirsman & Geeraerts, 2006, p. 271). Langacker (2008, p. 69) goes along the same lines: "In a narrow sense, we can characterize metonymy as a shift in profile" (of an image schema).

I propose the following thesis, on the basis of the narrower notion of domains presented in this chapter:

Thesis about metaphors and metonymies: Metaphors refer to mappings between domains; metonymies refer to meronomic and other relations within domains.

The upshot is that by narrowing the use of the notion of a domain, we obtain a sharper division between metaphors and metonymies. One theory that comes close—in content if not in terminology—is Croft's (2003). He starts out from the broad notion of domain in the Langacker tradition; but then, following Langacker (1987, p. 152), he introduces a distinction between (base) domain and domain matrix. Croft writes: "The combination of domains simultaneously presupposed by a concept such as [HUMAN BEING] [is] called a *domain matrix*" (2003, p. 168). In line with the proposed thesis, Croft describes metaphors as domain mappings (p. 177). He then argues that metonymy is based on what he calls *domain highlighting*, involving a shift from the foregrounding or highlighting of something in the primary domain of a concept to the foregrounding or highlighting of it in a secondary domain within the same domain matrix (p. 179). On the surface, this characterization differs from my notion of metonymy, but the result of Croft's analysis is basically the same as mine. That said, I believe that the contrast between dimensional domain mappings and meronomic relations makes the difference between the functions of metaphors and metonymies clearer.

2.6.2 The Contiguity Theory of Metonymy

The question arises whether all types of metonymy can be explained by a single cognitive mechanism. Peirsman and Geeraerts (2006) argue the implausibility of any unitary theory being able to do so. Before cognitive linguistics introduced the notions of domain and domain matrix, *contiguity* was the standard explanation for metonymy. Peirsman and Geeraerts consider whether contiguity might serve as an alternative to the mapping theory of cognitive linguistics. Given that contiguity suffers from the same vagueness as the notion of domain, they make an addition to its traditional definition, arguing that any theory of contiguity must be prototype based. Their definition of contiguity is conceptual in nature; they regard metonymy as something that is not objectively given but rather the result of different construal operations.

Peirsman and Geeraerts (2006) define spatial part-whole contiguity as the prototype for the category. Of course, this is the most typical meronomic relation; thus far, their proposal agrees with the one presented here. They argue that, on the next level of typicality, one finds container-contained relations such as "Oscar drank a glass too many" and "the milk tipped over." Next are location-located relations such as "the whole house woke up" or the use of "billiards" for "a room where billiards is played." The people who wake up are *not* parts of the house; rather, their locations are parts of the house's location. In this weaker sense, container-contained and location-located relations can be seen as meronomic relations—again, in accordance with my thesis. According to Peirsman and Geeraerts, the least typical form of spatial metonymy is adjacency: for example, the use of "old wig" to mean "old person." In summary, I find little conflict between the approach of Peirsman and Geeraerts (2006) and the one I propose here. The main difference is their focus on how metonymies can be more or less prototypical.

This and the previous section have shown the central role of dimensional domains for a cognitive theory of semantics. Once the domains are given, concepts can be defined in relation to them. For this, however, an account of how concepts are learned must be presented.

2.7   Learning in Conceptual Spaces

We are not born with our concepts; they must be learned. I must therefore account for how the relevant regions of a conceptual space are created from the experience of the agent. To be useful, the concepts must not only be applicable to known cases but should generalize to new situations as well.

The cost of generality is the increase of error. Learning concepts can be more or less successful. When a particular perception is sorted under a concept, this may be a miscategorization, which in turn leads the agent to choose the wrong action. For example, a person may sort a particular berry under the concept *edible* and end up with stomach problems or worse. If she realizes that she has made a mistake, she will adjust the application rules for the concept that led to the error, and thereby hopefully avoid the same mistake in the future.

Learning a concept often proceeds by generalizing from a limited number of exemplars of the concept (see, e.g., Reed, 1972; Nosofsky, 1986, 1988; Langley, 1996). Adopting the idea that concepts have prototypes, we can assume that a typical instance of the concept is extracted from these exemplars. If the exemplars are described as points in a conceptual space, a simple rule that can be employed for calculating the prototype from a class of exemplars is that the position of the point representing the prototype is defined to be the *mean* of the positions for all the exemplars (Langley, 1996, p. 99). The prototypes defined in this way can then be used to generate a Voronoi tessellation. Applying this rule means that a prototype is not assumed to be given a priori in any way but is completely determined by the experience of the subject. Figure 2.8 shows how a set of nine exemplars (represented as differently filled circles), grouped into three categories, generates three prototypical points by calculating means (represented as black Xs) in the space. These prototypes then determine a Voronoi tessellation of the space.

The mechanism illustrated here shows how, once the structure of a domain is established, the application of concepts can be generalized on

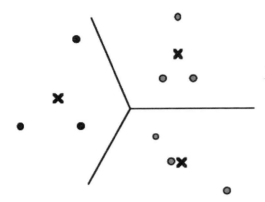

Figure 2.8
Voronoi tessellation generated by three classes of exemplars. The Xs represent the prototypes.

the basis of only a few examples of each concept. The additional information that is required for the generalization is extracted from the geometric structure of the underlying conceptual space. In this way, conceptual spaces add information to what is given by experience.

Furthermore, the concepts generated by such a categorization mechanism are *dynamic* in the sense that when the agent observes a new item in a category, the prototype for that category will, in general, change somewhat, since the mean of the class of examples will normally change. Figure 2.9 shows how the categorization in figure 2.8 is changed after learning about one new exemplar, marked by an arrow, belonging to one of the categories. This addition shifts the prototype of that category, which is defined as the mean of the exemplars, and consequently the Voronoi tessellation is changed. The old tessellation is marked by hatched lines, and the old prototype is marked by a gray X. Categories bordering on the changed prototype will then also change their meaning, but categories that have no common border will remain the same. Thus meaning change that is caused by learning will be local.

Note that the learning mechanism outlined here leads to extremely quick learning. Given only one or a couple of examples of a category, an approximate prototype can be calculated. This partially explains how word learning can be so fast. If a new example of a category is observed, the previous number of observations of a category will determine how much the prototype should be adjusted. The larger the number of previous observations, the smaller the effect of a new observation will become. The

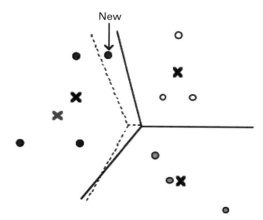

Figure 2.9
Change of Voronoi tessellation in figure 2.8 after learning a new exemplar.

mechanism thus explains how a language learner can quickly pick up an approximation of the meaning of a word, but full mastery of the meaning requires several encounters.

The proposed mechanism also explains some aspects of the *overgeneralization* that often occurs in concept formation. For example, when a child learns the word *dog*, the child will apply it not only to dogs but also to goats, horses, calves, and so on. The reason is presumably that the child masters only a few prototypes in animal space, and these prototypes will be used to generate a partitioning of the entire space. Consequently the child will overgeneralize a concept in comparison to its standard use. After all, communication functions better if the child has some word for any animal than if she has no word at all. However, when the child learns new prototypes for other animal concepts, she will gradually adjust an early concept to its normal use because her partitioning of the animal space will become finer when new prototypes are added.

2.8 The Vagueness of Concepts

Natural language is replete with vague terms.[35] When is somebody bald? Is this dog yellow or brown? Philosophers since Leibniz have dreamed of constructing a precise language where all vagueness is eliminated, where "every misunderstanding should be nothing more than a miscalculation," and where it would suffice for scientists "to take their pencils in their hands, to sit down to their slates, and to say to each other . . . let us calculate." Vagueness is, however, not a bug but a design feature of natural language. I will argue that there are good reasons related to cognitive economy why language contains vague terms.

First of all, note that there is no conflict between vagueness and the requirement that concepts be represented by convex regions. What convexity requires in relation to vague concepts is that if two object locations x_1 and x_2 both satisfy a certain membership criterion—for example, they have a certain degree of membership—then all objects between x_1 and x_2 also satisfy the criterion (Gärdenfors, 2000, sec. 3.5).

To a large extent, the vagueness of concepts is a result of the fact that we learn concepts by examples and counterexamples. The model of concept formation based on Voronoi tessellations that was presented in the previous section provides good clues to the mechanisms of vagueness. A first clue is that if prototypes are learned from the examples that have been encountered, the location of the prototype will be expected to move over time. Consequently the cognitive representation of how a prototype of a concept is located in a conceptual space may not be very precise (Douven, Decock, Dietz, & Egré, 2011). Figure 2.10 contains two closely located prototypes (marked by dots) for each of four concepts. They generate four different dividing lines between each pair of concepts. More generally, a probability distribution of the location of a prototype would, by the same mechanism,

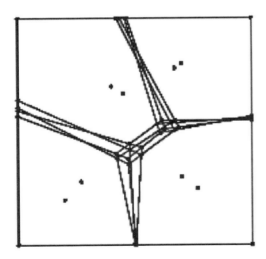

Figure 2.10
Multiple prototypes generate vague borders in Voronoi tessellations. Reprinted from Douven et al. (2011) with kind permission from Springer Science and Business Media.

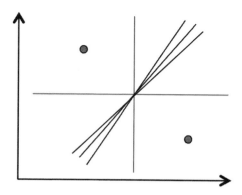

Figure 2.11
Changes in dimensional weights generate vague borders in Voronoi tessellations.

generate a distribution of dividing lines. For any point in the space, such a distribution can then be used to determine the degree of membership in a category. In this way, vagueness phenomena arise naturally.

Another clue to vagueness is that the relative weights of the dimensions in a domain are not precisely determined. For one thing, the weights often vary with the context. For example, depending on the relative weights of the two dimensions depicted in figure 2.11, the slope of the line dividing the space between the two prototypes (marked by dots) will vary. Again, a probability distribution over the weights will generate a distribution of dividing lines.[36]

In practical categorizations, both of the above mechanisms may have an influence on the vagueness of the borders between concepts.

The two mechanisms also generate a cognitive mechanism for phenomena related to *categorical perception* (e.g., Harnad, 1987, 2005; Petitot, 1989). This means that the effects of a categorization are amplified by the perceptual systems, so that distances within a category are perceived as being smaller and distances between categories are perceived as larger than they are according to the values of some physical dimension. For example, when somebody sings slightly out of tune, we tend to hear the tones as more correct than they are if the frequencies are measured and compared to the ideal values. When moving from one prototype toward another there will first be a region where the points are clearly classified with the first prototype, then a region where membership is rather quickly shifted toward the second prototype, and then a region where the points are clearly classified with the second prototype (fig. 2.12). The category membership function

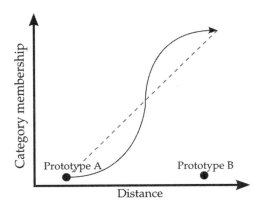

Figure 2.12
How vague membership distributions generate categorical perception.

thus exhibits an S shape that is characteristic of categorical perception. In brief, this model shows that membership in categories does not decrease linearly with distance but makes a quick jump from one category to another.

The upshot is that cognitive limitations concerning the locations of prototypes and the relative weights of dimensions explain why concepts in general are vague and why categorical perception is ubiquitous.

Although language is vague, linguistic mechanisms can be employed to counteract vagueness. In particular, *compositions* of concepts can serve to increase the precision of communication. For example, words like "very," "quite," and "almost" can be combined with vague adjectives like "bald" and "yellow" to create more precise expressions. The mechanisms underlying compositions of meaning will be the topic of chapter 13.

2.9 Comparisons with Some Other Approaches to Identifying Concepts

Several theories of concepts show similarities to the theory of conceptual spaces. In this section, I compare my theory with four of the most well-known.

2.9.1 Feature Analysis

In the philosophical tradition of Fregean logic and Tarskian truth conditions, it has been assumed that the meaning of a word can be decomposed into a finite set of conditions that are jointly necessary and sufficient to describe the word's meaning. The semantic analysis based on *distinctive features* goes one step further and also assumes that the necessary and sufficient conditions can be formulated in terms of a finite number of semantic primitives (cf. Jackendoff, 1983, p. 112; Goddard and Wierzbicka, 1994).[37] In linguistics and cognitive science, these assumptions have been expressed most prominently by Fodor and Katz (1963), R. Lakoff (1971), Schank (1975), and Miller and Johnson-Laird (1976) (although Miller and Johnson-Laird express some reservations).

Could it not be that the dimensions of a conceptual space can be generated from such a set of distinctive features? A number of experimental findings speak in favor of dimensional representations based on similarities in contrast to feature representations. The prototype theory of Rosch (1978) and her followers builds on data showing that objects can be more or less typical examples of a category, and there is graded membership in a category. These findings are difficult to explain in a theory of distinctive features. Furthermore, Smith and Medin (1981, pp. 121–123) present three types of results that go against feature analysis: (i) people can make fine discriminations about, for example, the size of objects, which implies that they have access to dimensionalized knowledge about the corresponding concepts; (ii) multidimensional scaling analyses consistently reveal dimensional properties; and (iii) in perceptual categorizations, as in Labov's (1973) experiment with containerlike objects, subjects distinguish, for example, diameter-to-height ratios that are used in their categorizations. Such ratios presume dimensional representations.

Much of the early work in cognitive linguistics is based on a contrast with a semantics based on necessary and sufficient conditions. Lakoff (1987) spends several chapters on this point. Langacker (1987, p. 54) criticizes feature analysis. He argues that "a cognitive domain is an integrated conceptualization in its own right, not a feature bundle." Also, Jackendoff (1983, pp. 112–122) expresses criticism. One of his examples is the concept *red*. *Red* must include the feature *color*, "but once the marker COLOR is removed from the reading of 'red,' what is left to decompose further? How can one make sense of redness minus coloration?" (p. 113). Thus Jackendoff, like Langacker, ends up proposing irreducible cognitive domains as a basis for semantics.

If the analysis is extended to features that have a dimensional structure, the position gets closer to the theory of *frames* and to the theory of conceptual spaces.

2.9.2 Frame Theories

The kind of representation proposed in the thesis about concepts is on the surface similar to *frames* with slots for different features that have been popular within cognitive science, linguistics, and computer science.[38]

In brief, the main difference between these theories and the one presented here is that I put greater emphasis on the *geometric structure* of the concept representations. For example, features in frames are often represented in a symbolic form. As will be seen in the following chapters, the geometric structures are cardinal for the analysis of many types of concepts and for combinations of concepts.

The definition of a concept presented here is richer than a frame representation, since conceptual spaces make it possible to talk about concepts being *close* to each other and about objects being more or less *central* representatives of a concept. The model that I propose can be seen as combining frames with prototype theory, although the structure of the conceptual spaces will make possible predictions that can be made neither in frame theory nor in prototype theory.

In linguistics, Fillmore's (1968, 1976, 1982) *frame semantics* is the best-known use of frames. It originates with case grammar (Fillmore, 1968): a case frame is described as "characterizing a small abstract 'scene' or 'situation,' so that to understand the semantic structure of the verb it was necessary to understand the properties of such schematized scenes" (Fillmore, 1982, p. 115). An example is the commercial transaction frame, the elements of which include buyer, seller, goods, and money. When explaining the semantics of verbs such as "buy," "sell," "pay," "charge," and so on, different elements of the frame are exploited. As we shall see in chapter 10, ideas similar to frame theory turn up in the conceptual space analysis of the meanings of verbs. The difference is again that by highlighting the geometric structure of the relevant domains, one obtains a richer way of modeling various semantic aspects.

Pustejovsky's (1991) theory of the *generative lexicon* can also be seen as a kind of frame theory. He introduces a "qualia structure" as a cross-categorical representational tool. The qualia are inspired by Aristotle's "modes of explanation" for entities and relations. Pustejovsky's qualia come in four types:

(i) *formal*: that which distinguishes it within a larger domain (typically its physical properties);
(ii) *constitutive*: the relation between an entity and its constituent parts (meronomic relations);
(iii) *telic*: the purpose and function of the entity, if it exists (typically why it was made);
(iv) *agentive*: factors involved in its origin (what brought the entity about).

The formal part can be seen as a parallel to the domain matrix of a concept (discussed further in chap. 6), and the constitutive part corresponds to the meronomic relations (also discussed in chap. 6). The telic part corresponds to functional properties that will be treated in section 8.4. There is, however, no correspondence to the agentive part in this book. The main reason for this is that what brings an entity about rarely has any influence on how it is categorized.

Pustejovsky's qualia structures thus share some similarities with the notions developed in this book. Once again, however, the qualia structures and the theory of the generative lexicon do not consider the underlying geometry of meanings.

2.9.3 Semantic Maps

In recent years, semantic maps have become popular among linguists (e.g., Haspelmath, 1997, 2003; Croft, 2001; Zwarts, 2010a; Sanso, 2010).[39] Like conceptual spaces, semantic maps are spatial representations of how a set of meanings "hang together," so that closeness on the map means that meanings are similar. The maps have mainly concerned grammatical expressions (parts of language belonging to closed classes), but they can also be applied to open class content words. The purpose of constructing a map is often typological in that the maps are supposed to represent meanings that are common to a group of languages or even universal to all languages. In other words, the hypothesis is that languages can differ in how they map the words from a particular domain onto a semantic map, but they cannot differ with respect to the underlying semantic map.

A good example is Haspelmath's (1997) map of indefinite pronouns (fig. 2.13), which is proposed as a universal map. He shows, for example, that the meaning of "somebody" in English covers (1)–(5); "nessuno" in Italian covers (4), (6), and (7); "dhipote" in Greek covers (5), (8), and (9); and both "quelconque" in French and "irgendein" in German cover the meanings (2)–(6), (8), and (9). An explicit assumption of the methodology of semantic maps is that the meaning elements of the maps must be organized

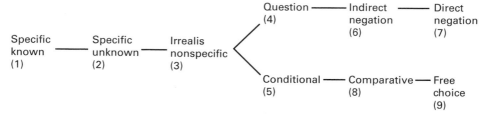

Figure 2.13
Haspelmath's (1997) semantic map of indefinite pronouns. Reprinted by permission of Oxford University Press.

so that, for each domain and for each language considered, the region corresponding to the meaning of a term must be *connected*. Croft (2001, p. 96) calls this the *semantic map connectivity hypothesis*.[40]

Zwarts (2010a) distinguishes between two approaches to semantic maps: one that he calls *matrix-driven*, and one that he calls *space-driven*.[41] Relations between meanings are derived either from cross-linguistic data or from semantic considerations. In the matrix-driven approach, which is the most common, the semantic map is "induced" in a data-driven fashion from the cross-linguistic lexical matrix.[42] In general, the semantic map is presented in the form of a graph. A point of this approach is that initially no assumptions are made about how these meanings hang together in the graph, but this is supposed to be the result of the inductive process.

However, one question that is seldom addressed within this approach is how the meaning components in the semantic map, such as the nine terms in figure 2.13, are *identified*; this is left to the intuitions of the linguist. The proposed functions are theoretical entities hanging in the air. Potentially, the theory of conceptual spaces offers a principled way of bringing order into this.

The space-driven approach builds on the assumption that, for a particular domain, one universal conceptual space provides the semantic structure for all natural languages. A prime example is the study of color terms that is based on the seminal work of Berlin and Kay (1969). For this domain, it is assumed that there exists a psychologically valid color space that is common to all people of the world, and thereby common to all semantic mappings (see sec. 2.2). This approach to semantic maps comes close to the one adopted in this book.

In brief, the dimensions of the semantic universe should be identified, and they should be related to the relevant communicative functions. An important requirement, however, is that meaning should be separate from form. This entails that the descriptions of semantic functions must be free of syntactic features. Instead they should be based on their grounding in perception and action and on their roles in communicative acts. In line with this, Croft (2001, p. 96) writes: "Not only should construction of the [semantic map] be empirically possible, but the structure of the [semantic map] should also make sense in semantic, pragmatic, and/or discourse-functional terms. The hypothesis of typological theory . . . is that most grammatical domains will yield universals of the form-function mapping that can be represented as a coherent [semantic map]."[43]

Haspelmath (2003) finds the same semantic functions in different maps (e.g., recipient and beneficiary) and mentions that they belong to a "semantic universe," but he does not develop the idea. In my opinion, it should be taken seriously (see Croft, 2001, chap. 2). The different semantic maps should be connected and unified so that we generate a picture of the entire semantic universe—the geography of the human mind (cf. Croft, 2003, p. 139).

In conclusion, this chapter has presented the basics of the theory of conceptual spaces. For more details, I refer the reader to Gärdenfors (2000). The notion of a domain has been reanalyzed in terms of dimensional structures. Domains form the core of the analysis of lexical semantics that will be presented in part 2 of the book.

3 The Development of Semantic Domains

When studying the semantics of a language, we should consider not only how words are mapped onto meanings. In chapter 1, I emphasized the epistemological criterion: a semantic theory shall explain how a language can be learned. Deacon (1997) and Kirby and Hurford (2002) argue that a language that survives over generations must be easy to learn by children. The epistemological criterion puts constraints on how the semantics of a language are structured. In other words, *how* words are learned matters for what they mean. In this chapter, I argue that sorting meanings into domains is a decisive factor in making language learning more efficient.

Child development research points to some evidence that certain domains are more basic than others. This applies in particular to domains that are *grounded* via sensorimotor programs. For example, Spelke and her collaborators (Spelke, 2000; Spelke & Kinzler, 2007) have identified four basic "core knowledge systems" representing (i) visuospatial structure, (ii) objects and their interactions, (iii) actions and goal directedness, and (iv) numbers and relationships of ordering. All but the fourth will be part of the list of meaning domains analyzed in this chapter.[1]

Other domains that children acquire derive from social interactions. For example, learning about kinship relationships cannot only be perceptually grounded. Another example is the domain of economic transactions: just observing coins and bills being exchanged for goods does not suffice to generate the domain of prices, loans, interest rates, and so on.

A consequence of the different character of domains is that the semantic learning mechanisms show some strong asymmetries. For example, why is it easier to explain to a four-year-old the meaning of the color terms "chartreuse" and "mauve" than to explain monetary terms like "inflation" or "mortgage"? The difference is not a matter of word frequency: the monetary terms are more frequent. Rather, the four-year-old masters the semantic *domain* of colors and thereby knows the meanings of many color words.

Adding new color terms is just a matter of learning the mapping between the new words and the color space: for example, "chartreuse" is a kind of yellowish green, and "mauve" is a pale violet. On the other hand, the child is normally not acquainted with the domain of economic transactions. Money for the four-year-old means concrete things—coins and bills—that one can exchange for other things. Abstract monetary concepts are not within her semantic reach. In line with this idea, Smiley and Huttenlocher (1995, p. 24) write: "Even a very few uses may enable the child to learn words if a particular concept is accessible. Conversely, even highly frequent and salient words may not be learned if the child is not yet capable of forming the concepts they encode."

The explanation I propose in this book is that our conceptual knowledge is organized into domains. This organization is necessary to make language learning possible. There is a difference in *cognitive effort*: grasping a new domain is a cognitively much more difficult step than adding new terms to an already established one.[2] Once a domain is common to a group of potential communicators, various means (words, gestures, icons, etc.) of referring to different regions of the domain can be developed. Conversely, if a domain is not shared, communication is hampered.

In this chapter, I present the most basic domains required for communication. The main thesis is that *a close parallel exists between the development of intersubjectivity and the development of semantic domains*. I will identify, and to some extent idealize, a number of domains that are central for the analysis in the rest of the book. I conclude the chapter by comparing the developmental analysis with some remarks on the evolution of semantics, something that is another source of constraints for how meanings are constituted.

3.1   Components of Intersubjectivity

One form of cognition that is well developed in humans, by comparison with other species, is intersubjectivity, which in this context means *the sharing and representing of others' mentality*. I take the term "mentality" here to involve not only beliefs but all forms of mental states, including emotions, desires, attentional foci, and intentions. In philosophical debate, intersubjectivity is commonly referred to as having a "theory of mind" (see, e.g., Premack & Woodruff, 1978; Tomasello, 1999; Gärdenfors, 2003, 2007c). I want to avoid this term, since it is often equated with representing the beliefs of others—something that, in the present account, is but one aspect of intersubjectivity.

The question of whether an animal or a child exhibits intersubjectivity does not have a simple yes-or-no answer (see e.g., Stern, 1985; Gärdenfors, 2003, 2007c; Brinck, 2008). Intersubjectivity will here be decomposed into five capacities: representing the emotions of others (empathy), representing the attention of others, representing the desires of others, representing the intentions of others, and representing the beliefs and knowledge of others. In Gärdenfors (2007c), I argued that this ordering is supported by both phylogenetic and ontogenetic evidence. In this section, I outline these different components of intersubjectivity as a basis for analyzing the semantic domains required and their role in the development of meaning.

Intersubjectivity is important to language learning because it adds information that helps the child interpret the meaning of an expression that is uttered (Bloom, 2000, 2002; Sperber & Wilson, 2002). Bloom (2002) argues that the forms of intersubjectivity relevant to communication are the same as those used in noncommunicative situation. He claims that the general capacity for intersubjectivity is a preadaptation for language learning.

Representing the Emotions of Others
The ability to share others' emotions is often called *empathy* (Preston & de Waal, 2003). Bodily and vocal expressions of emotions communicate the agent's negative or positive experiences. They are most obvious among social animals. Characteristically, facial signals, such as the play-face expressions of chimpanzees and gorillas, carry emotional rather than referential meaning. From a developmental point of view, empathy develops very early in human children (Stern, 1985).

Representing the Desires of Others
A desire is a positive attitude toward some external object or event. Understanding the desires of another individual requires understanding that the other may not like the same things that you do. In contrast, representing emotions concerns the inner state of an individual, without reference to an external object. Results from child development studies (e.g., Repacholi & Gopnik, 1997; Wellman & Liu, 2004) suggest that eighteen-month-old children can understand that others have desires other than those they have themselves.

Representing the Attention of Others
Representing the attention of others means that one can understand, for example, that someone else is attending to some particular object or event. Humans, other primates, some other mammals, and some birds are good

at gaze following (Emery, 2000; Kaminski, Riedel, Call, & Tomasello, 2005; Bugnyar, Stöwe, & Heinrich, 2004). Even very young children can understand where other people are looking. *Shared attention* results when two agents are simultaneously attending to the same target. A more sophisticated version is drawing *joint attention* to an object, where the agents also are aware of the attention of the other. I analyze the mechanism of joint attention in detail in the following chapter.

The achievement of joint attention is a necessary condition for intentional cooperation. The ability to engage in joint attention and reciprocal behavior vis-à-vis a third element has, so far, been clearly demonstrated only in humans (though for a different view see Gómez, 2007; Leavens, Hopkins, & Bard, 2005, 2008).

Representing the Goals and Intentions of Others

Because the human cognitive system takes self-motion as a cue for goal directedness, intentions to act can be read directly from observation of behavior (Michotte, 1963). More recently, Gergely and Csibra (2003) argue that infants primarily interpret instrumental actions not by their causality but by their efficiency. Notice that though one can interpret someone else's behavior as goal directed and can follow her gaze to a target, this need not mean that one represents her intention. It is sufficient to represent the goal of her action. Only later in their development do children adopt a mentalistic stance, learning to attribute intentions to the actor.[3]

There is a stronger sense in which intentionality may be shared, which is crucial for complex forms of cooperation. That concerns *joint intention* (Tomasello et al., 2005), where an individual represents the plans of another and coordinates her own intentions with the goals of the other.[4] Nonlinguistic examples are playing in a football team and in a string quartet. This requires that the agents (i) share an intention to interact, (ii) react to each other's individual intentions to act, and (iii) coordinate their respective intentions. Coordinating complementary roles toward a future goal that is not present in the shared context is even harder, because which action to take next will not be evident from the context.

Representing the Beliefs and Knowledge of Others

The most advanced test of intersubjectivity among humans and other animals involves finding out whether they represent what others believe or know. Tomasello and Call (2006) review the experimental evidence for chimpanzees; they conclude that the chimpanzees "know not only what others can and cannot see at the moment, but also what others have seen in the immediate past." This suggests that the apes have some limited representation of the information available to others.

It is easier to test whether young children can understand that "seeing is knowing," since one can communicate with them through language from a fairly early age. The most common method uses so-called *false-belief tests* (see, e.g., Perner, Leekam, & Wimmer, 1987; Gopnik & Astington, 1988; Mitchell, 1997). In a nonverbal version of the test (Call & Tomasello, 1999), children performed as well as in the verbal form, while all apes that were tested failed the test.

A correlation exists between language proficiency in children and their ability to pass the false-belief tasks (Astington & Jenkins, 1999). In particular, parental use of mental predicates in child-directed speech correlates with children's performance in false-belief tasks (de Villiers & Pyers, 1997). As a consequence of the dependency on language, Southgate, Senju, and Csibra (2007) shifted to experiments using eye tracking instead of verbal responses. With this methodology, it could be established that fifteen-month-olds can already understand the beliefs of others (Onishi & Baillargeon, 2005).[5]

Humans often know what someone else knows: that is, they have second-order knowledge. They can, however, also have higher-order knowledge and belief. Using irony in communication, for example, involves intersubjectivity at the fifth level: when I say, ironically, "He's so smart," I believe that you believe that I want you to know that I do *not* believe what I am saying. Similarly, politeness also involves five levels (Gómez, 1994): When I say, "Can you pass the salt?" I believe that you believe that I want you to know that I want the salt. The capacity for higher-order beliefs forms the basis of *joint beliefs*, which are collectively often called *common knowledge* (H. Clark, 1992, chap. 1; Stalnaker, 2002).

The five components of intersubjectivity I have outlined here are exploited so naturally in adult human communication that their importance evades us. That they appear successively in children's development and are present to a lesser extent in nonhuman cognition suggests that they have played an important evolutionary role. My thesis is that the more advanced components have coevolved with our communicative capacities; they prepared the ground for new forms of human cooperation (Gärdenfors, Brinck, & Osvath, 2012).

3.2   Domains in Semantic Development

Using conceptual spaces as my modeling framework, I next trace the development of semantic knowledge in children by identifying and describing the domains that are required for different levels of intersubjectivity and for various basic forms of communication.

I must emphasize that it is not so much a process of addition as a *separation out* of domains from the child's original "blooming buzzing confusion" (James, 1890) that contains information from all domains simultaneously. What happens developmentally is that one domain after the other is separated out and can be attended to, indeed, as a separable set of dimensions. For example, two-year-olds can represent whole objects, but they cannot reason about the dimensions of those objects. Goldstone and Barsalou (1998, p. 252) note:

Evidence suggests that dimensions that are easily separated by adults, such as the brightness and size of a square, are treated as fused together for children. . . . For example, children have difficulty identifying whether two objects differ on their brightness or size even though they can easily see that they differ in some way. Both differentiation and dimensionalization occur throughout one's lifetime.

An example mentioned earlier is Piaget's (1972) studies of "conservation." Children younger than five years cannot separate the volume of a liquid from its height. When choosing between two glasses of lemonade, they pick the glass with the highest level of lemonade even though that glass is very narrow and the other is wide. If the lemonade is poured into a shallow container, they believe there is less of it. Only later do they learn that the volume of a liquid is conserved between containers and not always correlated with height. In other words, they have learned to separate the dimension of height from that of volume. In this way, their conceptual domain is expanded with a new domain.

That I call certain sets of dimensions domains means that I assume that these dimensions are separable. In most cases, I know of little empirical evidence that I can use to support these assumptions. For many perceptual domains, the relevant dimensions can be identified, though their structure is not completely known. Establishing that they are separable from those of other domains would involve extensive psychophysical work. Nevertheless, given the analysis in section 2.5, I believe that my use of the notion of domain is more precise than what is common in linguistics.

3.2.1   Emotion Domain

The importance of empathy in human (and animal) interaction highlights the question of how emotions are represented mentally. Several competing theories about the structure of the emotion domain exist. As mentioned earlier, most of these theories contain two basic dimensions: a *value*

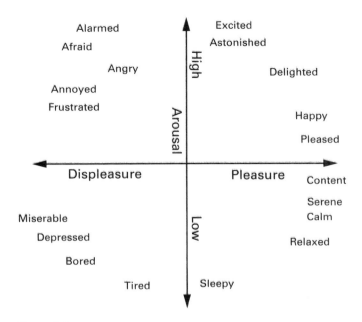

Figure 3.1
A two-dimensional emotion space (from Russell, 1980).

dimension on a scale from positive to negative, and an *arousal* dimension on a scale from calm to excited emotional states (e.g., Osgood, Suci, & Tannenbaum, 1957; Russell, 1980).[6] The product space of these two dimensions allows a spatial representation of the basic emotions (fig. 3.1). Distances in the space indicate degrees of similarity between emotions.

It is well known that emotive intersubjectivity is an important aspect of mother-infant "attunement" interactions (Stern, 1985). The infant learns the correlations between different emotions and the corresponding facial and vocal expressions. In other words, the child learns how to map behaviors into an emotion space. Sharing an emotion means that the participants in the exchange are in emotional states that are closely located within the same emotion space; that is, the emotions are "attuned." This is arguably the most fundamental way of sharing meaning.[7]

Animal calls and other forms of signals are mainly tied to emotions: the vocalizations signal danger, food, or a mate or create contact between mother and infant. The warning calls of vervet monkeys may primarily be not warnings but expressions of fear, which then secondarily have obtained the function of warning the other troop members (Goodall, 1986, p. 125).[8]

The upshot is that both evolutionarily and developmentally, the sharing of the emotional domain has deep roots. Individuals who share emotions can coordinate their actions, for example, by attacking or fleeing together. The communication involved here need be no more than the nonintentional expression of emotions.

In human communication, emotions are conveyed by intonation patterns, rather than words. Fernald (1992) has studied how mothers and infants in different cultures communicate. She found four basic, more or less universal, intonation patterns: (i) the mother *encourages* the infant to do something ("can you get it?") by raising her pitch; (ii) she *rewards* the child for something the child has done ("good girl") by lowering her pitch; (iii) she *warns* or *forbids* the child, in which case the phrasing is often staccato ("no, no, no"); and (iv) she *comforts* the child, in which case the speech melody shows a soft, billowing pattern ("poor, poor Eddy").[9] The main function of this type of communication is to *share* emotions.

3.2.2 Visuospatial Domains

During her first months, the child learns to coordinate her sensory input—vision, hearing, and touch—with her motor activities (Thelen & Smith, 1994). This generates a narrow, egocentric space that basically maps onto her visual field. The subsequent role of this space in intersubjective engagement is manifested, for example, by the child's ability, from six months of age, to follow the gaze direction of her mother, if the mother turns her head to look at an object within the visual field of the child (D'Entremont, 2000). From twelve months, the child can follow her mother's gaze if the mother just turns her eyes toward the object (Butterworth & Jarret, 1991). Representing the attention of others means that one can understand when someone is looking at some object or noticing some event. As suggested here, even very young children can understand where other people are looking.

The visual domain expands throughout the child's development. A few months later, a child can follow the gaze of others even if they look at points outside her immediate visual field. This requires that the represented visual space extend beyond the current visual field to cover the entire physical space. The child can now understand references outside her visual field. It should be understood that the represented physical space is not just an extension of the visual domain but an *amodal* abstraction from visual, auditory, tactile, and perhaps even olfactory perceptions.[10]

A more advanced transformation of the represented space comes with the ability to represent an *allocentric* space, that is, a space as seen from another point of view (Piaget, 1954).[11] This involves a shift of perspective. A concrete example of the use of allocentric space is the ability to direct someone whose vision is blocked.

More precisely, the visuospatial domains should be seen as a combination of an allocentric representation of physical space with an egocentric representation provided by the visual system. The two representations are connected with two different types of uses: the egocentric for reaching and interacting with objects, the allocentric for navigating through the environment (Gallistel, 1990). This double aspect of physical space is revealed by the two linguistic codes we have established for referring to positions: egocentric *left* and *right*, and allocentric *west* and *east* (or *north* and *south*).[12] Similarly, what is *behind* the house from my egocentric perspective may be *in front of* the house from an allocentric perspective (see sec. 11.4).

3.2.3   Force and Action Domains

Research in cognitive science and in cognitive linguistics has tended to focus on vision and spatial cognition. Throughout the book, I argue that not only space but also *forces* are important in how we represent the physical world. Experiments on how humans perceive the movement of persons and other objects suggest that the kinematics of movement contain sufficient information to identify the underlying *dynamic force patterns* (e.g., Johansson, 1973; Giese & Lappe, 2002; Giese & Poggio, 2003). This capacity seems to develop in early infancy (White, 1995). Leslie (1994) speculates that infants have a domain-specific module that calculates the movements of objects according to force dynamics (he calls it the Theory of Body Mechanism). Children also use the actions performed by an object as a cue to its categorization, in addition to shape and other static domains (see, e.g., L. Smith, 2005). Thus the force domain can be understood as a shared domain for purposes of communication. Note that forces as represented in our minds are psychological constructs and not Newton's scientific concept.[13]

In chapter 8, I analyze actions in terms of the forces involved in generating those actions. The basic premise is that an action can be represented as a pattern of force vectors. The force pattern for running is different from the force pattern for walking, and the force pattern for saluting is different from that of throwing.

As we shall see, forces are also ubiquitous in metaphorical constructions. For example, language often describes applications of mental or social force, such as when one person is threatening or persuading another. In such cases, the term "power" is often substituted for "force" (Winter & Gärdenfors, 1995; Gärdenfors, 2007a).

3.2.4 Object Category Space

Objects are not only located in physical space but also represented in an *object category* space with its own quality dimensions (Gärdenfors, 2000, chap. 4). If the visuospatial domain represents *where* an object is, the object category space represents *what* it is.[14] The object category space is composed of a number of separable domains: color, size, shape, weight, temperature, and so on. Thus the object category space is a kind of metadomain.

Although communicative coordination in the emotion and physical domains can be achieved without words, coordination in the object category space is strongly enhanced by the use of words. The first fifty or so words acquired by children consist mainly of object category words for perceptually identifiable concrete objects: people, food, body parts, clothing, animals, vehicles, toys, and household objects (Fenson et al., 1994). They are often used in situations involving joint attention between the child and an adult.

Hurford (2007, p. 224) notes that declarative pointing only communicates the location of an object and indicates nothing about its properties. This means that pointing may function without a shared object category space having been established. In a situation of joint attention, parents often scaffold children with words to provide information about a category domain related to the object of attention. This means that the communication takes place in the product space of the visuospatial domain and the object category space.

As Goldin-Meadow (2007) and others have demonstrated, children combine pointing gestures with words before they rely on words alone. The words complement pointing or gaze sharing and thus expand the possibilities for shared meaning domains in the communicative situation. The minds of the communicators meet in two ways: in the visuospatial domain and in the object category space. Only later does the child learn words for more abstract domains, such as kinship relations or money.

During the period between eighteen and twenty-four months, children undergo what is called a naming spurt, acquiring a substantial number of nouns for representing objects. Evidence suggests that, during this period, they also learn to extract the general shape of objects, and this abstraction improves object category learning (Son, Smith, & Goldstone, 2008; L. Smith, 2009). How the object category space develops in children is still not well known. Some cues can be obtained from children's ability to learn nonsense words for new things (Bloom, 2000; L. Smith, 2009). There seems to be a *shape bias*; that is, the shape of objects seems to be the most important property in determining category membership for small children (Landau, Smith, & Jones, 1998; Smith & Samuelson, 2006). As already mentioned, children also overgeneralize concepts (MacWhinney, 1987; Bloom, 2000).

3.2.5 Value Domain

Understanding that others may not have the same desires as oneself requires a representation of a *value domain*, one that is separated from other domains.[15] This capacity develops before the ability to represent the beliefs of others (cf. Flavell, Flavell, Green, & Moses, 1990; Wellman & Liu, 2004), emerging somewhere between fourteen and eighteen months (Repacholi & Gopnik, 1997). Whereas emotions express how an individual feels, desires express an individual's attitudes toward objects, events, and other agents. Because desires are *relational*, representing the desires of others is more cognitively demanding than representing their emotions. One frequently used method to model an individual's value domain is to introduce a utility function that assigns values to objects or situations. Other representations exist, but I will not pursue the structure of the value domain here, since it does not play a major role in the semantic theory I am presenting.

A reasonable hypothesis given the empirical data, however, is that children initially consider the value of an object to be *intrinsic* to the object, that is, a domain of the object category space, such as weight or size. Only later is the value domain separated from the category domain, so that different individuals may be represented as assigning different values to the same object.[16]

3.2.6 Goal and Intention Domains

A human throws at the future
an arrow tied to a string.
The arrow lands in an *image*,
and the human reels itself in towards that object.
—Paul Valéry

I turn next to how goals and intentions can be represented in conceptual spaces. First, note that any representation of intentions requires that goals already be represented: the goal domain is therefore primary and must be described first. For simplicity's sake, consider a situation where an agent is located at a certain physical distance from a desired object. In this case, the goal domain can be constructed from the physical domain: reaching the goal is reaching the object. The difference between the two domains is that, in the physical domain, the locations of agents and objects are in focus, while in the goal domain, the focus is on the *distances* between them. In this example, the goal domain is the space of movement vectors from the initial to the desired location. Mathematically this can be described as the product space of physical space with itself. A goal is then a vector $<a,o>$, where a is the location of the agent, and o the location of the desired object. The goal is achieved when $a = o$, that is, when the agent and the object are at the same place.

When the goal is represented in this way, two principal ways of obtaining the goal present themselves. One is that the agent moves to the goal location and grasps the object. The other is that the agent communicates, for example via imperative pointing (see chap. 4), so that another individual brings the object to the agent.

In general, goal vectors can be more abstract than movement vectors in the physical domain: in principle, they can be defined in all kinds of semantic domains. If I want the wall to be painted purple, my goal is to change its color from the current location in the green part of the color domain to the desired location in the purple subregion. The goal is now represented as a vector in the color domain. In economics, cognitive science, and artificial intelligence, goal spaces are represented as abstract spaces. The classic example from artificial intelligence is Newell and Simon's (1972) General Problem Solver. These goal spaces can be viewed as generated by metaphorical extensions from the original physical space, always maintaining the key notion of distance. This analysis is supported by the pervasiveness of spatial metaphors in relation to goals: "he *reached* his goal," "the goal was *unattainable*," "the target was set *too high*," and so on (Lakoff & Johnson, 1980).

Consider next the problem of representing intentions in conceptual spaces. My basic analysis is that the intention domain is a product space of the goal domain and the action domain. An intention is thus a combination of a goal and an imagined action conceived of as leading toward that goal.[17] I present the action domain in chapter 8, and in chapter 10 I show that intentions play an important role in the semantics of verbs.[18]

Children talk about their own intentions earlier than about those of others. By two years, children use a subject term, usually "I," together with a result verb (see sec. 10.4) such as "go," "open," or "eat," to indicate what they are about to do, that is, their intention (Huttenlocher, Smiley, & Charney, 1983, p. 90). Corresponding constructions for the intentions of others appear considerably later in their development.

3.2.7 Age and Time Domains

During their second year, children learn to use the words *old* and *young*, which indicate that they have appropriated the *age dimension*. They can use this dimension to classify people. However, an understanding of an abstract time domain develops much later. Spatial metaphors for time are ubiquitous: it is difficult to talk about time without using words that originate from the visuospatial domain. Piaget (1927/1969, p. 43) emphasizes that "time and space form an inseparable whole" in children's minds. In other words, young children cannot treat space and time as separate domains. This situation parallels the dimensions of height and volume, which are inseparable for preschool children but become separable when children learn that volume is a conservative dimension. Similarly, space and time start as a single metric that becomes gradually differentiated into two separable domains.[19]

In general, children use spatial terms earlier than temporal terms (E. Clark, 1973). At five years, they can use a large time vocabulary. Nevertheless children do not develop a separated representation of temporal relations until they are nine or ten (Harner, 1981). Casasanto, Fotakopoulou, & Boroditsky, (2010) show that children's representations of time depend more on space than representations of space depend on time. They conclude that kindergarten children can ignore temporal information when making spatial judgments, but they have difficulty ignoring spatial information when making temporal judgments.

As I argue in chapter 10, an explicit representation of time is not needed for understanding verbs. Hence children can correctly use verbs even before they have developed a separate time dimension.

3.2.8 Event Domain

What is involved semantically in representing the beliefs of others, for example, knowing that someone has a false belief? Beliefs are normally expressed in terms of *sentences*. So how are the meanings of sentences related to semantic domains? One possibility, which I defend in chapter 9, is that most simple sentences express *events* (I include *states* among events). There I model an event in terms of two vectors: a *force vector* that typically represents an action (performed by an agent), and a *result vector* that describes a change in the physical (movement) or an object category (change of property) domain of a "patient." Consequently the event domain is cognitively more complex than other domains.

Given this model, one may reasonably speculate that understanding the beliefs of others requires understanding their representation of events. If this is correct, it is no wonder that understanding the beliefs of others develops rather late in childhood. For example, Nelson (1996) shows how the use of the word "know" develops over time in children and does not achieve its ordinary meaning until after children can pass the false-belief test (see fig. 3.5).

I have now identified a number of semantic domains that are required for children's communication. Several domains are based on the levels of intersubjectivity that appear in their development. I have outlined how these domains can be represented with the aid of conceptual spaces. Since independent semantic evidence suggests that the domains are necessary for modeling basic meanings, their connection to intersubjectivity can be used as a stepping-stone to an analysis of the development of semantic knowledge. I return to the role of the domains in the semantic analyses in part 2 of the book.

3.3   Correlations to Development of Language

A central thesis of this book is that the domains form a central part of semantic knowledge. In this section, I present some linguistic evidence that the development of semantic knowledge can appropriately be described as an increasing set of separable domains.

In the analysis of child language data, the establishment of a word in a child's vocabulary is often analyzed in terms of the average frequency of the word's usage at a certain age. Typically the frequency of a word's usage starts at or close to zero, increases rapidly, and then levels off once the word is established in the vocabulary.[20] The resulting curve thus has an S shape. I will call the interval when usage increases most rapidly the *establishment period* for a word.

I can now formulate a general thesis concerning semantic domains:

Establishment thesis: If one word from a domain is learned during a certain establishment period, then other (common) words from the same domain should be learned during roughly the same period.

Clark (1972, p. 753) proposes a similar hypothesis, although she writes about semantic fields rather than domains. She hypothesizes that "children form semantic fields even before they learn the complete meaning of words, and that, therefore, they will make substitutions among those terms that have features of meaning in common." To test the establishment thesis, I have analyzed data from the CHILDES corpus. The publicly available Web-based ChildFreq application is a highly efficient tool for such investigations.[21] Here I can present only a few examples from my analysis.

For most of the domains that I discussed in the previous section, words are established during the language spurt that takes place between twelve and twenty-four months. This holds true in particular for the different domains of the category space. For example, consider the domain of fruits. Figure 3.2 shows the frequency curves for the names of several of the most common fruits: apple, banana, pear, grapefruit, and orange. As the diagram shows, these words have an establishment period between twelve and eighteen months of age. "Orange" has a longer period of increase, presumably because the word is also used within the color domain. The two meanings of the word cannot be separated in the ChildFreq data.

In some domains, the words are clearly established later. One such example is the domain relating to life and death. Figure 3.3 shows that the

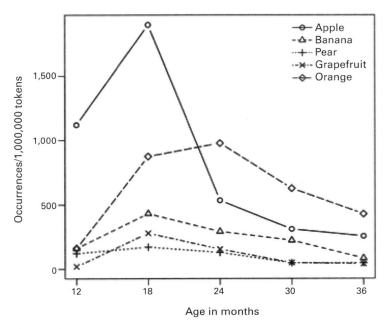

Figure 3.2
The establishment periods for common words for fruits.

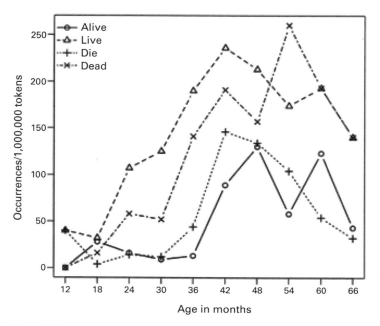

Figure 3.3
The establishment periods for some words in the life domain.

establishment of the words "live," "die," "alive," and "dead" occurs, for the most part, between thirty and forty-two months.

Another example is the domain relating to knowledge and memory. Figure 3.4 shows the frequency curves for the words "believe," "remember," "forget," and "guess." In this case, the establishment period occurs between thirty-six and fifty-four months. Note that these words concern an individual's relation to facts, and in that way they relate to the event domain discussed earlier. Furthermore, the period coincides with the period when children learn to pass the false-belief tests that are based on linguistic feedback.

The next example from ChildFreq concerns the levels of intersubjectivity that I discussed in section 3.1. It is difficult to find a clear correspondence between these levels and the learning of particular words. Anyway, I have chosen the verb "look" as an indicator of understanding the attention of others; "want to" and "wanna" as indicators of understanding desires; "going to" and "gonna" as indicators of understanding goals and intentions; and "know," "think," and "believe" (the latter two combined into one category) as indicators of understanding belief and knowledge.[22]

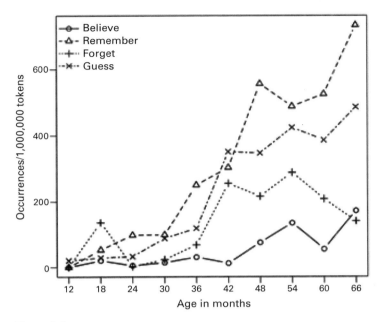

Figure 3.4
The establishment periods for some words in the knowledge domain.

In accordance with the analysis presented here, Gopnik and Meltzoff (1997, p. 121) write:

The emergence of belief words like "know" and "think" during the fourth year of life, after "see," is well established. In this case . . . changes in the children's spontaneous extensions of these terms parallel changes in their predictions and explanations. The developing theory of mind is apparent both in semantic change and in conceptual change.

Figure 3.5 suggests that the order of the establishment periods conforms to the one I proposed in section 3.1. A more detailed analysis of the uses of these words, in different contexts, is required to more clearly establish the connection with intersubjectivity. Note that "know" and "think/ believe" do not quite follow the usual S shape. This may partly be explained by the many idiomatic uses of these words, which make their usage frequencies grow more continuously.[23]

Although I have here only presented a limited number of examples, it should be clear that the establishment thesis is rich in empirically testable predictions.[24] ChildFreq is a limited but quick-to-use tool for testing the thesis. For more in-depth analyses, more sophisticated methods from

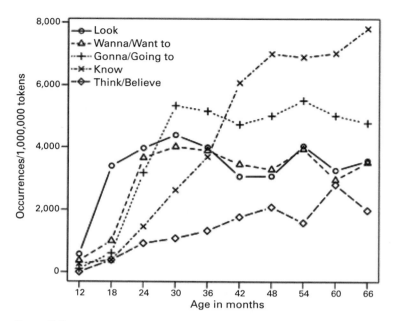

Figure 3.5
The establishment periods for some verbs related to stages of intersubjectivity.

corpus linguistics should be employed. I invite linguists and child development researchers to test the establishment thesis.

That metaphors do not come alone offers further evidence for the
domain organization of semantic knowledge. Lakoff and Johnson (1980)
argue that metaphors are organized around themes such as "argument is
war," "time is a resource," and "more is up." In the previous chapter, I
proposed that a metaphor expresses an identity in topological or geometric
structure between different domains.[25] That is, a word that represents a
particular structure in one domain can be used as a metaphor to express
the same structure in another domain. Once a metaphor has established
such a mapping, it can be exploited to generate other metaphors from the
same domain.

An example of such a mapping is the designation of certain computer
programs as "viruses." This metaphor from the biological domain created
a new way of looking at this class of programs. It suddenly opened up
possibilities for expressions like *invasive* viruses, *vaccination* programs, and
hard disk *disinfection*. Again, the establishment of metaphors should be
further analyzed on the basis of corpus data.

This concludes my presentation of the main conceptual domains that early language learning is based on. There is, however, much more to be done in investigating the development of the use of domains in children's language. In chapter 7, I return to a special application in relation to how adjectives are learned. Methods from developmental psychology should also be applied to investigate how children appropriate new domains.

3.4 Evolution of Semantics

Quand les hommes ne peuvent changer les choses, ils changent les mots.[26]
—Jean Jaurès

Why is *Homo sapiens* the only species that has a symbolic language? Why does language have words as building blocks that can be composed in ever new ways to create ever new meanings? As a complement to a description of how different domains develop in children's semantic spaces, I briefly bring up the issue of how the human meaning system has evolved. I do not subscribe to any strong form of parallelism between evolution and development. Nevertheless evolutionary considerations can be a source of inspiration for constraints on meaning systems and for ideas on which semantic structures are the most fundamental.

In evolutionarily early forms of communication, the communicative *act* in itself and the context it occurs in were presumably more important than the expressive form of the act (H. Clark, 1992; Winter, 1998; Gärdenfors, 2010). As a consequence, the *pragmatics* of natural language are the most fundamental from an evolutionary point of view. When communicative acts become more varied and eventually conventionalized during hominin evolution, and their contents become detached from the immediate context, one can start attending to the expanding meanings of the acts. Then semantics become salient. Finally, when linguistic communication becomes even more conventionalized and combinatorially richer, certain markers (i.e., syntax) are used to disambiguate the contents when the context and the common ground of the interlocutors are not sufficient. According to this view, syntax is required only for the most subtle aspects of communication—pragmatic and semantic features are more fundamental for communication.

This view of the evolutionary order of different linguistic functions stands in sharp contrast to mainstream contemporary linguistics. For followers of the Chomskian school, syntax is the primary study object of linguistics; semantic features are added when grammar is not enough; and pragmatics are a wastebasket for what is left over (context, deixis, etc.). However, I believe that when the goal is to develop a theory of the evolution of communication, the converse order—pragmatics before semantics before syntax—is more appropriate.[27] In other words, there is much to find out about the evolution of communication before we can understand the evolution of semantics, let alone the evolution of syntax.

The pragmatics of a communication system will not determine its semantic structure. There are many ways to carve up a meaning space; color terminology is a good example here. Nor will the semantics of a language determine its syntax. However, the semantic structures will induce constraints on what syntactic constructions are possible or likely; the semantic maps presented in section 2.9.3 are examples of such constraints.

The historical changes of a language are difficult to explain. The meanings of words and the syntax of a language change through time as whimsically as a leaf falling from a tree. It is impossible to predict these more or less random movements. However, for the falling leaf, its general direction is determined by gravitation. Newton's laws present constraints on movements. Similarly, the goal of a semantic theory is to formulate corresponding laws for meaning changes, for which meaning structures are more likely than others and for how the structures constrain syntax. The aim of this book is to argue that conceptual space is a fruitful framework for modeling meaning structures and their dynamics.

In support of the position that pragmatics is evolutionarily primary, it is clear that most human cognitive functions had been chiseled out by evolution before the advent of language. Language would not be possible without these cognitive capacities, in particular having rich intersubjectivity, having a memory system that includes episodic memory, and being able to represent future goals (see Gärdenfors, 2003, 2004c, 2008; Gärdenfors et al., 2012). In contrast, some researchers argue that human thinking cannot exist in its full sense without language (e.g., Dennett, 1991). According to the latter view, the emergence of language is a cause of many forms of thinking, such as concept formation. It is true that many concepts, for example, *inflation, month, heir*, are probably unattainable without language. However, this does not contradict the assertion that the majority of our concepts are established, via perception or action, before they find linguistic expression.

However, seeing language as a cause of human thinking is like seeing money as a cause of human economics (Tomasello, 1999, p. 94). As long as humans have existed, they have been trading goods. But when a monetary system does emerge, it makes economic transactions more effi-cient. The same applies to language: hominins have been communicating since long before they had a language, but language makes the exchange of meanings more effective.

The analogy carries further: when money is introduced in a society, a relatively stable system of *prices* emerges. Similarly, when linguistic com-munication develops, individuals will come to share a relatively stable system of *meanings*, that is, components in their conceptual spaces, which communicators can exchange between each other. In game-theoretical terms, meanings are *equilibrium points* in a system of exchanges just as prices are. In this way, language fosters a *common structure* of the mental spaces of the individuals in a society: it leads to a meeting of minds.

I have argued elsewhere (Gärdenfors, 2003, 2004c, 2007c; Gärdenfors et al., 2012) that the evolution of meaning is best explained by assuming a coevolution of intersubjectivity, cooperation, and communication. Along similar lines, my aim here is to show that *sharing meaning in different domains makes possible various new forms of cooperation*. I take this as the central selective advantage of sharing semantic domains.

Sharing the visual domain facilitates coordination, which can be achieved without explicit communication, for example, by gaze following.[28] Stron-ger forms of cooperation become possible when the participants achieve *joint attention* toward an object or a place. Joint attention can be established simply by means of mutual gazes, but pointing is a more efficient way to communicate the mutual goal. I analyze this process in detail in the fol-lowing chapter.

For many forms of cooperation among other animals, it seems that sophisticated mental representations are not needed. If the common goal—for example, a prey—is present in the immediate environment, the collabo-rators can focus on it directly before acting (Boesch & Boesch-Achermann, 2000). If, on the other hand, the goal is distant in time or space, then a *joint* mental representation of it must be available before cooperative action can be taken.

Anthropologists generally agree that hominins evolved in open land-scapes that favored a long-ranging lifestyle (Preuschoft & Witte, 1991; Hilton & Meldrum, 2004). To survive in this type of environment, it became important to be able to jointly refer to objects that were not present on the scene. Language functions not just for directing attention but also for coordinating it (Richardson & Dale, 2005; Tylén et al., 2010). The possibility of achieving joint attention to absent entities via commu-nication opens up new forms of cooperation—in particular toward future states, involving coordinated action toward a nonpresent common goal. This introduces selective pressures toward a communicative system that makes it possible to *share mental representations of nonpresent entities.*

Planning for future collaboration, essentially a task of coordinating goals, requires coordination in the physical domain (often outside the visual field), the category domain, the action domain, and the goal domain. (As I have shown earlier, these are domains that develop early in children.) Bickerton (2009) argues that scavenging large fauna is the crucial form of cooperation in the evolution of hominins. They needed ways to communicate what (e.g., a carcass) had been found, where it was located, and how their scavenging actions might be coordinated. Such planning depends on forming joint intentions, an advanced form of intersubjectivity presumably unique to humans (Tomasello et al., 2005). A *joint plan* can be described as a combination of a joint intention and coordinating actions.

Iconic communication, in the form of miming (Donald, 1991), can achieve much of this coordination, certainly for the emotion, visuospatial, and action domains. However, miming about object category domains is much more difficult. In that case, arbitrary symbols are more efficient for the task. Along the same lines, Tylén et al. (2010, p. 6) write:

Analogous to the way that manual tool use has been shown to enlarge the peripersonal space by extending the bodily action potential of arm and hand in space . . . linguistic symbols liberate human interactions from the temporal and spatial immediacy of face-to-face and bodily coordination and thus radically expand the *interaction space.*

I submit that the evolution of symbolic communication about shared physical, object category, action, and goal domains generated evolutionary advantages for the individuals of a society built around cooperation toward future goals.

In human society, many forms of cooperation are based on *conventions* (Lewis, 1969). The central cognitive requirement concerning conventions is that they presuppose enduring *joint beliefs* or common knowledge. For example, if two cars meet on a narrow gravel road in Uganda, then both drivers know that this coordination problem has been solved numerous times before by keeping left; both *know* that both know this; both know that both know that both know this; and so on. The result is that they both keep to the left without any hesitation. Although conventions may be established without explicit communication, communication makes the presence of a convention clearer.

Joint beliefs form the basis of much of human culture. Commitments and contracts are special cases of joint-belief-involving cooperation about the future. When you commit yourself to an action, you intend to perform the action in the future; the person you commit to wants you to do it and intends to check that you do it; and there are joint beliefs concerning these intentions and desires (Dunin-Kepliz & Verbrugge, 2001; Jackendoff, 2007, chap. 11). Where no joint beliefs exist, commitments cannot arise. It seems that children as young as three already have some understanding of joint commitments (Gräfenhain, Behne, Carpenter, & Tomasello, 2009). In particular, joint beliefs are necessary for *semantic conventions*: when I use a word, I assume that my interlocutor has a joint belief concerning its meaning. Humpty Dumpty misses this point when he claims that "when *I* use a word, it means just what I choose it to mean—neither more nor less."

In summary, my position is that intersubjectivity and the sharing of the corresponding semantic domains together constitute the decisive characteristics of human cooperation, and these mechanisms form the evolutionary background for increasing communicative capacities, leading to symbolic language.

4 Pointing as Meeting of Minds

4.1 Introduction

In this chapter, I present an analysis of the development of the semantic domains of a child from gesture to verbal communication.[1] As a driving example, I analyze the different forms of pointing. One reason for focusing on this topic is that pointing is a prelinguistic universal of human communication (Liszkowski et al., 2011). I aim to show that the meaning processes involved in communication by pointing are essentially the same as those in spoken communication, and the development of the linguistic communication ability can be seen as a transition from pointing in the visuospatial domain to "pointing" in other domains. These domains include the emotion domain, the object category space, and the goal domain. Using combinations (products) of the visuospatial domain with other domains, I argue that there is a semantic continuum in development, and purely verbal communication arises from a bootstrapping process grounded in gestural communication.[2]

The semantic framework is based on "meetings of minds" that I develop further in the next chapter. According to this framework, the meanings of expressions do not reside in the world or solely in the mental schemes of individual users but develop via communicative interactions (cf. Brinck, 2001, 2004b). A meeting of minds occurs in pointing when the interactors experience an alignment of their attention in the visuospatial domain, and in verbal communication when the interactors experience an alignment of their attention in other conceptual domains (H. Clark, 1992; Pickering & Garrod, 2004).

Pointing is a special gesture that serves as an interface between the physical environment and the semantic domains of the communicators (Brinck, 2004a, 2008). Although two interlocutors can establish joint attention just by following each other's gazes, pointing is a more efficient tool for this. Pointing is often used in conjunction with words, and it plays an important role in the development of verbal language in children (McNeill, 1992; Goldin-Meadow, 2007). Not only do different types of pointing activities serve different communicative purposes, but they also differ in terms of their cognitive representation, which I will model in terms of conceptual spaces.[3]

Another key thesis of this chapter is that growing semantic complexity is achieved by expanding the domains of the shared conceptual space. The basic operation that is used to compose multiple domains can be modeled as creating product spaces. The expansion by composition of domains generates a continuum of communication situations. In this way, the spatial approach provides an underpinning for the developmental sequence of gestural and verbal communication.

I use a classification of different kinds of pointing basically borrowed from Bates, Camaioni, and Volterra (1975), but with some further refinements from Brinck (2004a), Tomasello, Carpenter, and Liszkowski (2007), Goldin-Meadow (2007), and Gärdenfors and Warglien (2013). I call the individual doing the pointing the *pointer* and the onlooker the *attendant*. I am not concerned with the exact timing of the different forms of pointing in child development. My objective is rather to reconstruct the semantic developmental continuum. However, I do draw extensively on the existing empirical evidence. The nature of the domains that are introduced also affects the type of intersubjectivity that is involved, as presented in the previous chapter.

4.2 Imperative Pointing

Since Bates et al. (1975) distinguished between imperative and declarative pointing, the imperative form has been recognized as the most elementary. It is performed to make the attendant *do* something for the pointer (e.g., bring the toy the pointer is pointing to). In this type of pointing, the pointer treats the attendant as a causal agent that can be influenced by pointing (Bates, 1976; Brinck, 2004a). In principle, imperative pointing is therefore not necessarily an intentional act of communication; it could be like pushing a button that triggers a chain of causal events. As a pointer, you can learn to point without considering other agents, for example, as a mere result of reinforcement learning. Often, however, imperative pointing has communicative intent. Infants who point imperatively often monitor the attention of their social partners.

Figure 4.1
Imperative pointing. P = pointer; A = attendant; Obj = object. The thick line is the direction of pointing. Thin lines are related to the attendant's gaze. The dashed line from the pointer to the object is the closure of the triangulation performed by the attendant.

Cognitively, the only factor that needs to concern the pointer is her egocentric visuospatial domain in which the focal object is located. Thus from the pointer's view, no intersubjectivity is necessarily involved. This conclusion is supported by Tomasello's (1999) observation that in this stage of development, children can master pointing without understanding the pointing of others. However, the attendant must understand the desire of the pointer. The attendant must also identify the location of the object with the aid of the direction of the pointing and stopping at the first object in that direction that fits with the desires of the pointer (fig. 4.1). If no other clue is given, the first object found is chosen.

4.3 Declarative Pointing

Declarative pointing involves directing the attention of the attendant toward a focal object (Bates et al., 1975; Brinck, 2004a; Tomasello et al., 2007). It consists of one individual pointing to an object or spatial location and at the same time checking that the attendant is focusing his or her attention on the same object or place (Bates, 1976; Brinck, 2004b). The attendant in turn must check that the pointer notices that the attendant is attending to the right entity. Such joint attention is a good, but fallible, mechanism for checking that the minds of the interactors meet in focusing on the same entity (Tomasello, 1999; Tomasello et al., 2007). The crucial difference with respect to imperative pointing is that the child need not desire to obtain the object pointed at, but rather desire to achieve joint attention to the object with the attendant (Brinck, 2004a).

For reasons of pragmatics, I distinguish three types of declarative pointing: *emotive, information requesting*, and *goal directed*.[4] In emotive declarative pointing, the pointer wants the attendant to communicate her emotions concerning the object. In information-requesting pointing, the pointer wants nonemotive information about the object, for example, information about the linguistic label for the object. Finally, in goal-directed declarative pointing, the pointer helps the attendant achieve a goal.

4.3.1 Emotive Declarative Pointing

Some authors claim that the emotive form where pointing has an evaluative use is the more fundamental (Brinck, 2001). For example, the child points to an object that she finds frightening to obtain a reaction of fear or reassurance from the attendant. The main benefit of this kind of exchange for the child is that she can learn about objects vicariously. This primary function presupposes that the child can understand the emotions and the attention of the addressee, but it does not require the understanding of intentions or beliefs (Brinck, 2008). Sharing emotional reactions is exciting in itself for an infant. In many cases, infants engage in emotive pointing simply because they want to interact with the adult. In such cases, the object pointed to is just a tool for sharing experiences.

In addition to the visuospatial domain involved in imperative pointing, emotive declarative pointing obviously builds on the emotion domain. Minimally, emotive declarative pointing thus takes place in the product space of the visual and the emotion domains. Emotive declarative pointing requires both domains to be available for the participants. Adding the emotion in a pointing situation enriches the context. In turn, the visuospatial domain involved in pointing may enable the alignment of the emotions of the participants.

In emotive declarative pointing, both participants must check the attention of the other, and at least the pointer must check the emotional state of the other (fig. 4.2). If successful, emotive declarative pointing entails the convergence of the participants in two domains: the egocentric visuospatial domain (gazes converge on the same object) and the emotion domain (both express compatible emotions).

Children react if the attendant does not show an expected emotional response. In an experiment by Liszkowski, Carpenter, and Tomasello

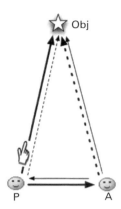

Figure 4.2
Emotive declarative pointing. P = pointer; A = attendant; Obj = object. In addition
to what is involved in figure 4.1, the pointer and attendant converge in their emotional spaces, reaching a joint emotion.

(2007), an adult correctly identified what the child was pointing at but, in
different conditions, expressed either interest or disinterest in the object.
In the disinterest case, which is a mismatch in the emotion domain, the
child rapidly decreased pointing activities (as compared with the interest
condition). The infant lost interest because the other person did not engage
in a meeting of minds. Another experiment involving infant pointers
shows that if the attendant responds with the correct emotion but simply
does not look at the focal object, the infant expresses disappointment,
since the adult does not engage in interaction (Liszkowski et al., 2004).

4.3.2 Information-Requesting Pointing

Information-requesting pointing has the purpose of receiving nonevalua-
tive information about the object pointed at. A typical example is when a
child wants to know the word for the relevant object category. In this type
of pointing, a child often accompanies the gesture with "da" (short for
"what's that?") or some similar protodemonstrative.

Information-requesting pointing typically combines the visuospatial
domain with the object category space. A new word uttered by the attendant indicates to the pointer that the object belongs to a category that is
not unfamiliar to her, and it may trigger a further partitioning of the object
category space.

All acts of declarative pointing involve intentional communication and
can be seen as proto–speech acts (in line with the speech acts studied by

Searle, 1969). Both emotive pointing and information-requesting pointing are forms of proto*interrogative* speech acts (information-requesting pointing more explicitly so). And to the extent that it is an intentional form of communication, imperative pointing is, indeed, a proto*imperative* speech act.

4.3.3 Goal-Directed Pointing

Goal-directed declarative pointing can be introduced by an example from Liszkowski et al. (2007). A child observes an adult searching for an object that has been misplaced (e.g., a pair of glasses) and shows him the object by pointing. This kind of pointing supports the fulfillment of the attendant's goal. The pointer understands the goal from the actions of the attendant and perceives the mismatch between the current state and the goal and points so as to help the attendant achieve the goal. The result is that the pointer gives the attendant sufficient information to solve the coordination problem. In terms of speech acts, this type of pointing can be seen as a proto*declarative*. In the example, another solution would be that the child brings the object to the adult. It is important to note that the intersubjectivity of the pointer requires understanding only the *goals* of the attendant, not his beliefs (Brinck, 2001, 2004a).

Recognizing the actions of someone else seems to be a fundamental cognitive process. Going from understanding actions to understanding the goals behind them is not automatic. However, it develops during the first year in children (Tomasello, 1999). In other words, the intersubjectivity of goals is present when children begin to point in a goal-directed manner.

The goal domain supplies a natural representation of goals. In the case of pointing to an object, the problem is generated by a mismatch between the attendant's location and the object location, together with the attendant's lack of awareness concerning the location of the object. Pointing is triggered by the difference between the attendant's desired state and the current state. In this case, the goal domain is derived from the visuospatial domain: to reach a goal is to reach a location. The perceived goal is that agent and object should be at the same point. Pointing solves the problem by helping the attendant to move to the object. Of course, an alternative (but more costly for the pointer) fulfillment of the goal is to bring the object to the attendant.

In emotive declarative pointing, the visuospatial domain is composed with the emotion domain to help determine the meaning of the pointing gesture. In goal-directed declarative pointing, it is the product of the visuospatial domain with the goal domain that determines the meaning of the pointing. More elaborate forms of declarative pointing can be derived from combinations of these primitive forms. For example, pointing can be a form of inquiry to get evaluative information (is this food good?) about objects related to the goals of the pointer (Brinck, 2001, 2004a).

Some researchers claim that infants understand and produce imperative pointing before declarative pointing (e.g., Camaioni et al., 2004), but others argue that both types appear at the same time, that is, at twelve months (Carpenter, Nageli, & Tomasello, 1998; Liszkowski et al., 2004). At any rate, declarative pointing, in contrast to imperative pointing, is produced with the goal of achieving joint attention, and therefore the pointer must have some understanding of the attention of the other. For example, if the pointer indicates an object that does not fulfill the attendant's goals, the attendant should signal disappointment (no meeting of minds has occurred).

Convergence in declarative pointing is both meeting of eyes and meeting of minds. In the emotive case, participants have to find resonating emotional states. Formally, this can be described as convergence in the product of the visuospatial domain and the emotion domain. As I have argued, meeting of eyes but misalignment of emotions is disappointing for the pointer.

Whether a pointing gesture is emotive, information requesting, or goal directed may by itself be indeterminate; what type it is must be decided by context (Brinck, 2004a). The context is generated by gazes, gestures, posture, and emotive expressions. Which kind of "common ground" (H. Clark, 1996; Pickering & Garrod, 2004) is available to the pointer and attendant will depend on which product of domains becomes more prominent.

To achieve joint attention, the attendant is required to locate the first salient object along the pointed direction. A problem is that there may be several objects along that line, and one must be chosen. In the absence of other contextual criteria, the most obvious is the first one encountered. But, of course, knowledge of the context may suggest the selection of another object. In an emotive context, for example, when the pointer shows surprise, the selected object should be new to the pointer. In this case, the pointing is emotive declarative. In goal-directed situations, the actions of the attendant generate a context including a goal that determines which object is in focus. This argument shows why it is necessary to compose the visuospatial domain with other domains to resolve ambiguities.

Even if the participants correctly represent the perceptions of each other's visual space, the shape of the space itself may obstruct or create mismatches in the triangulation procedure. For example, obstacles can create nonconvexities in the visual space: the salient object may be blocked from the view of the attendant. If all points between the participants and the focal object are visible for both participants, the problem can always be solved. This indicates the interactive importance of the convexity of the visual fields (Warglien & Gärdenfors, 2013). What is shared is the intersection of the visual fields. It constitutes a special case of the fixpoint construction that I examine in the next chapter.

4.4   Declarative Pointing Composed with Words

When pointing is not selecting focal objects uniquely, it is helpful to compose the visuospatial domain with further domains (emotion domain, object category, or goal domain). Verbal language dramatically expands the possibilities of multiplying spaces. I now examine some of these mechanisms.

In addition to the emotive and goal domains, words trigger richer semantic domains. Objects not only constitute points in the visuospatial domain but are also represented in the object category space. In chapter 6, I propose that a noun refers to a concept in a multidimensional region in the object category space. By using a noun in connection with pointing, the visuospatial domain is composed with such a category domain. Pointing and noun constrain each other: pointing indicates a linear region of the visuospatial domain (where), while the noun indicates a region in the object category space (what) that determines a subset of the objects available on the scene. The joint constraint makes it easier to identify the focal object. For example, a small boy was pointing at two neighboring objects, a toy and a saxophone, saying "guitar," which was his overgeneralized noun for all music instruments. Actually a true guitar was located a few steps away, but pointing was discriminating between the two instruments, while the word was selecting between the two neighboring objects.

If a noun is not sufficient to select a unique focal object along the line of pointing, an adjective may do the job. According to Gärdenfors (2000), an adjective refers to a region of a domain in a conceptual space (I develop this idea in chap. 7). Saying "the red one" while pointing to similar objects close in space (e.g., a number of books) may define a unique solution that a combination of pointing and a noun could not solve.

Because the physical world is crowded with objects, pointing may often be indeterminate. I have now presented several examples that have a common pattern: by composing the visuospatial domain with different types of domains (emotion, goal, object category), pointing becomes a multidimensional activity that facilitates the selection of a unique reference object. We not only point in the visuospatial domain but learn to simultaneously point in our other semantic domains. Facial expressions point into the emotion domain, actions point into the goal domain, and words point into the object category space (and other domains). In this sense, one can say that words point to concepts.[5]

4.5 Language without Pointing

When the interactors are communicating about the external world, pointing is sufficient to make minds meet on a referent. However, when they need to share referents in their inner worlds, a different tool is required. In a sense, language is a tool for reaching joint attention by "pointing" to places in our inner worlds. This mechanism is bootstrapped by pointing, other forms of gestures, and emotive expressions. Goldin-Meadow (2007, p. 741) goes beyond this metaphorical assertion and writes that in children, "pointing gestures form the platform on which linguistic communication rests and thus lay the groundwork for later language learning." For example, prelinguistic children at about twelve months can sometimes refer metonymically to an absent object by pointing to a place where that object has recently been or is normally located (Liszkowski et al., 2007; Liszkowski, Schäfer, Carpenter, & Tomasello, 2009). For example, an infant may point to Daddy's chair as a reference to the absent daddy. This type of pointing can be seen as a transition from using pointing as a proto-demonstrative with a temporary meaning (see sec. 12.4) to using pointing symbolically (Daddy's chair being a symbol for Daddy) with a more permanent meaning (cf. sec. 1.5.2).

It can be argued that children's early gestures, in particular their declarative pointing gestures, provide the ground for later language learning. Support for this claim comes from a study by Iverson and Goldin-Meadow (2005), who found a clear correlation between the appearance of pointing and the development of spoken language. In brief, the earlier children begin to point, the earlier they start using language. The study also showed that the appearance of gestures together with words corresponds to the age when children start using two-word utterances. These findings also support the general claim that a continuity exists between gesturing and speaking in the development of communication.

Goldin-Meadow (2007, p. 742) notes that "mothers often 'translate' their children's gestures into words, thus providing timely models for how one- and two-word ideas can be expressed in English." Learning a word enables the child to make a *projection* (a dimensional reduction) from the product of the visuospatial domain and the object category space to the object category space alone.[6] This projection reduces the redundancy created by the mother's translation of gestures into words. My interpretation is that the mothers (and others) scaffold the developmental sequence in which children start communicating about the visuospatial domain, then learn to use the object category space in combination with the visuospatial domain to make communication more effective, and, finally, make it possible to detach themselves from the visuospatial domain by projection on the object category space. Like Wittgenstein, the child throws away the visuospatial ladder when it is no longer needed.

In this way, communication becomes detached from the current environment (Hockett, 1960). Language then opens up for new fields of communication: future cooperative plans, absent persons (gossip), imaginary entities and situations (play language and storytelling), and so on. Once this level of representation is reached, the roles and functions of communication change drastically. Words (rather than fingers) are mainly used to point to one's inner world, hoping that the listener can view a similar place in her inner world, from her perspective.

Communication is a matter of how minds meet. In my view, the processes of creating joint referents (and other meanings) are essentially the same in pointing and in speaking (cf. Brinck, 2001, 2004b, 2008). Traces of this mental pointing can be seen in the metaphors we ordinarily use to speak about communication, for example, "Do you see what I mean?" and "Do you follow me?"

While reference in the object category space can lose connections to the visuospatial domain, as in narrative, it still resorts to pointing as the basic mechanism for achieving a meeting of minds. In fact, pointing gestures are frequently reintroduced in story telling and other detached uses of language. Their function now is to give a visual complement to what the words point to in other domains (Haviland, 2000; McNeill, 2005, p. 40). Bühler (1982) calls this "deixis at phantasma."[7] Again a product space is created, but now it is the visuospatial domain that is added. Haviland (2000) provides several interesting examples of this phenomenon.

One concerns a Tzotzil speaker who tells a story about returning to a place where he had left a dying horse. When he says, "It was getting late," he looks up at the place in the sky where the sun would have been at that time. This is a deictic gesture that metaphorically describes the time of the event in the story. Deixis at phantasma reinforces the claim that verbal language and pointing gestures are conjoined in a unique semantic structure.

4.6 The Development of Pointing

The development from simple imperative pointing to the sophisticated process of directing others' attention in inner semantic domains can be seen as a process that builds and combines the visuospatial domain with other domains of growing complexity and dimensionality, generating multiple levels of mutual understanding. The additional domains I have considered in this chapter are the emotive and goal domain and the object category space.

To sum up the analysis, the developmental sequence of pointing can be described as an expanding set of product spaces:

1. *Imperative pointing.* Only the mapping of the visuospatial domain is implied. The pointer need not have any communicative intention (but often has).

2. *Emotive declarative pointing.* The visuospatial domain is combined with the emotion domain. Communicative intent is present.

3. *Information-requesting pointing.* The visuospatial domain is combined with the object category domains. Communicative intent involves an expectation that the attendant can provide information about the object pointed at.

4. *Goal-directed declarative pointing.* The visuospatial domain is combined with the goal domain. The communicative intent here implies also a representation of the other's goals.

5. *Pointing together with words.* The visuospatial domain is combined with the object category domains, the action domain, and other domains. In this case, the product of domains has to be coordinated with a combination of communication modalities (visual plus auditory).

6. *Detached language.* In this case, communication is based on the object category domains and other domains without combining it with the visuospatial domain. Communication aims at pointing in the other's inner space.

I have shown how products of the visuospatial domain and other domains provide a basic framework to understand where the different kinds of communication are located along a semantic continuum, and how purely verbal communication may arise from a bootstrapping process grounded in gestural communication. The theory represents in a single semantic framework gestural as well as verbal communication.

While my analysis has essentially aimed at establishing the semantic continuity of pointing gestures and verbal communication, we can also outline a *pragmatic* account (cf. Brinck, 2004a). Pointing acts can fruitfully be analyzed as analogous to speech acts in line with Searle's (1969, 1983) analysis.[8]

0. Grasping.   This is a direct action resulting in control of an object. The primary goal is to use the object (e.g., for sucking, for eating, or for placing it somewhere). Then this action can develop into a secondary goal of evaluating or learning about the object. Grasping is not primarily a communicative act. However, if the object is grasped for showing, then it becomes communicative.

1. Imperative pointing.   Instead of direct grasping, the pointing act, if successful, has the result that someone else performs the action and brings the object to the pointer with the same result as for grasping (Vygotsky, 1978, p. 56). This is a *protoimperative* form of communication. In the terminology of Searle (1983), the "direction of fit" of this communicative act is world-to-pointing: the act is supposed to bring about changes in the world, so that it matches the desire of the pointer.

2. Emotive declarative pointing.   This does not result in any grasping of the object, but if joint attention is established, then the pointer can achieve vicarious evaluation of the object via the emotional reaction of the attendant. This form of declarative pointing still involves an imperative element on the part of the pointer: Help me evaluate the object! This can be seen as a *protointerrogative* form of communication. Again, the "direction of fit" is world-to-pointing: now the act is supposed to bring about a reaction in the attendant that satisfies the pointer.

3. Information-requesting pointing.   If there is joint attention, the pointer can learn about the object via the gestural or linguistic reactions of the attendant. This form also involves a *protointerrogative* form on the part of the pointer: Inform me about the object! As in the previous case, the "direction of fit" is world-to-pointing.

4. Goal-directed declarative pointing.   This form reverses the roles of the pointer and the attendant. Here it is the attendant who wants to interact with the object (for using or for evaluating). The pointing helps the attendant to achieve this goal. Of course, the attendant may interact with the object at a distance (look at it, throw something at it), so grasping the object is not necessary. This form has no imperative component but is purely communicative. In Searle's (1983) terms, the "direction of fit" of this communicative act is pointing-to-world: the pointing is supposed to match the world so that, for instance, the attendant can find an object.

Goal-directed declarative pointing is a *protodeclarative* form of communication. We thus see that the correspondence to the three major forms of sentence moods appears already in communication by pointing.

5–6. Transition to verbal language. The use of words allows detachment of reference. Pointing acts turn into speech acts proper. As I have tried to show, however, the pragmatic mechanisms discussed by Searle (1969, 1983) have close parallels in pointing acts. Words can point not only to elements of the other's inner world but also to nonpresent or even nonexistent entities. This provides the ground for activities like counterfactual and strategic reasoning, prospective planning (Gärdenfors & Osvath, 2010), play language, and narratives.

The purpose of this chapter has not been to discuss gestural communication in general but to show that the meanings of different forms of pointing depend on sharing different kinds of conceptual domains. Other forms of gestures could, of course, be analyzed in similar ways.[9]

5   Meetings of Minds as Fixpoints

Of one and the same thought, from mind to mind, by what means—through what channel—can "conveyance" be made? To no other man's is the mind of any man immediately present. . . . "Under yon tree, in that hollow on the ground, lies an apple." At and during the time we are thus conversing, the ideas of the apple, the ground, and the hollow are in both our minds.
—Jeremy Bentham

5.1   Meanings as Emergent in Communication

In the previous chapter, I presented various forms of pointing as examples of how references to objects in the world can be established as a meeting of minds. I also showed how the referential process could be extended to meetings of minds in inner worlds. In this chapter, I generalize the analysis of communication to a model with some mathematical structure based on the notion of *fixpoints*.[1,2] Although most of the mathematics are relegated to an appendix, the text will be a bit more technical than in other chapters.

In relation to communication, Clark (1996, p. 19) distinguishes between *autonomous* and *participatory* actions. He writes: "Speaking and listening have traditionally been viewed as autonomous actions, like playing a piano solo." He argues that speaking and listening should be seen as participatory actions. Typically the goal is to make the minds of the communicators meet so that successful joint action can be taken.

An early version of a theory of meaning as a meeting of minds was formulated by Mead (1934).[3] He writes that meaning "arises and lies within the field of the relation between the gesture of a given human organism and the subsequent behavior of this organism as indicated to another organism by that gesture. If that gesture does so indicate to another organism the subsequent (or resultant) behavior of the given organism, then it has meaning" (1934, pp. 75–76).[4] Note that Mead uses "gesture" in a very general sense that includes all forms of vocal communication. He summarizes his position as follows: "Meaning as such, i.e., the object of thought, arises in experience through the individual stimulating himself to take the attitude of the other in the reaction to the object" (p. 89). A similar, more semantically oriented theory is proposed by Larsson (1997, 2008). He argues that "the ontological status of meaning is *intersubjective*, neither 'objective,' entirely grounded in the phenomena and entities of the world, nor 'subjective,' belonging to a single individual and thus private" (Larsson, 2008, p. 31; my translation).

Achieving joint attention is an example of reaching a fixpoint in communication. When my picture of what I point out to you agrees with my understanding of what you are attending to, my communicative intent is in equilibrium. Conversely, when what you attend to agrees with your understanding of what I want to point out to you, your understanding is in equilibrium. Two general features of this process should be noted. First, joint attention requires intersubjectivity in that both communicative partners can attend to the attention of the other. Second, the two partners need not have the same image of each other's inner state: it is perfectly possible for you and me to reach joint attention without my picture of your attention being aligned with your attention (I present examples of this in sec. 5.8). In other words, joint attention never requires transcending one's own subjective realm.

At the other end of the communicative spectrum, all reference is to the inner worlds of the partners. When a bedtime story is told about fictional characters in a fantasy world, there is no reference to anything in the environment; everything takes place in imagination. Nevertheless the storyteller and the listener most of the time succeed in coordinating their inner worlds to reach a mutual understanding of the story.

As long as communication is conceived as a process through which the mental state of one individual affects the mental state of another, then a "meeting of the minds" will be that condition in which both individuals find themselves in compatible states of mind, such that no further processing is required.[5] Just as bargainers shake hands after reaching agreement on the terms of a contract, so speakers reach a point at which both believe they have understood what they are talking about. Of course, they may actually mean different things, just as the bargainers might interpret the terms of the contract differently. It is enough that, in a given moment and context, speakers reach a point at which they *believe* there is mutual understanding. The ubiquity of "assent" signals in conversations (H. Clark, 1992;

Winter, 1998) nicely demonstrates the importance of the mutual awareness of such meetings of mind in everyday communication. In passing, it should be noted that this view clearly goes against the "conduit metaphor" of communication discussed by Reddy (1979). Before entering into the mathematical modeling, I will discuss the role of such signals in the following section.

In the tradition of cognitive semantics, image schemas are presented as structures that are common to all speakers of a language. Within the socio-cognitive type of semantics I present in this book, I do not assume that everyone has the same conceptual space, only that well-behaved mappings exist between the spaces of different individuals—well-behaved in the sense that the mappings have certain mathematical properties, to be specified in section 5.5.

In the present framework, minds meet when a function mapping states of mind onto states of mind, via language or some other form of communication, finds a fixpoint. Communication is like a dog on a leash pulling its master toward something the dog has in mind. The dog will pull, and its master will follow until an equilibrium is reached. The place where they stop is, literally, a fixpoint.

My approach takes significant inspiration from the communication games that have been studied by Lewis (1979), Stalnaker (1978), and others. To this foundation, I add assumptions about the topological and geometric structure of the various individuals' conceptual spaces to allow me to specify more substantially how the semantics emerge and what properties they have. Communicative acts can be understood as moves in such games. Not only may the players in a communication game have different interests and preferences, but they may also have different semantic spaces.

5.2   Levels of Communication

Wise men talk because they have something to say; fools talk because they have to say something.
—Unknown author

The obvious goes without saying. If all partners in a cooperating group perform their tasks as expected by the others, they have no need for communication. Cooperation takes place on the level of praxis. Only when an instruction or a correction is needed does communication play a pragmatic role. Language is mainly used to convey nonobvious features of our environment. The basic level of communication is therefore for solving problems of coordinating actions. However, there are situations when the communicators misunderstand each other, because of badly formulated instructions or because they have different mental models of the world. On this level, the communicators must coordinate their inner worlds. Coordination of inner worlds can also be done as a preparation for future collaboration.

There is also another, more severe form of misunderstanding that occurs because the addressee does not understand an expression used by the speaker, or does not understand it in the same way. For example, if you say, "I'll talk to the chair," and you mean the chairperson, while "chair" for me just means a physical object, I will not understand your intention. On this final level—the level of semantic coordination—the communicators must negotiate their use of expressions until they find a sufficient agreement.

To sum up, I want to distinguish three levels of communication, in addition to a ground level of human interaction:

Level 0: *Praxis.*   On this level, people interact with each other without using intentional communication.

Level 1: *Instruction.*   On this level, coordination of action is achieved by instruction.[6]

Level 2: *Coordination of inner worlds* (common ground).   On this level, people inform each other so as to reach a richer or better coordination. It can also be achieved via questions.

Level 3: *Coordination of meanings.*   On this highest level, people negotiate the meanings of words (labels) and other communicative elements.

A special case of level 3 arises when someone also exerts semantic power and dictates the meaning of a word. Specialists in different fields often have the power to determine the meaning of particular words, and everyone else will simply have to accept that the specialist knows what she is talking about when fixing a meaning (see, e.g., Putnam's [1975] discussion of "division of labor").

These four levels of coordination have been analyzed by Winter (1998), and I summarize them here in figure 5.1.

The four levels are used in a hierarchical manner. When one level does not function properly, a break is signaled, and the communication moves to the next higher level. When the problem is solved, the communicators signal an acknowledgment and return to the level below.

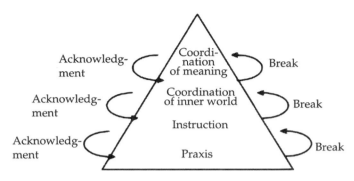

Figure 5.1
Four levels of communicative coordination. Based on Winter (1998).

Levels 1–3 can be matched with different types of failure of communication (Bara, 2010, pp. 193–194). If *A*, who has alcohol problems, asks *B* to bring the candy and *B* brings brandy, it can be failure on level 1 (mishearing) or failure on level 2 (not wanting to fulfill *A*'s intention), or semantic mismatch on level 3 (*B* thought *A* meant brandy, because that is an alcoholic's candy). When *A* signals the failure of communication, *B*'s reaction should make it clear what type of mistake is involved.[7]

Clark's (1992, 1996) work on *common ground* (see sec. 1.5.2) and *uptakes* can be seen as analyses of some central forms of coordination on level 2. First, the utterances in a conversation introduce new referents or new information about referents. This, together with the participants' expectations about the others' previous knowledge, forms the common ground that the subsequent conversation can take for granted. Second, a participant often introduces a proposal for a joint project in the conversation. This proposal can be taken up by the interlocutor (or it can be rejected). The proposal and the uptake then lead to a coordination of the continuing communication.

A related proposal is the *interactive alignment account,* proposed by Pickering and Garrod (2004). They divide the coordinating processes in a dialogue into phonetic, lexical, syntactic, semantic, and situation model alignments (2004, fig. 2, p. 176). The phonetic, lexical, and syntactic alignments do not have any direct correspondences in the model proposed here. Although their framework focuses on how a dialogue triggers activations on various linguistic levels in the interlocutors and not the coordination of meanings considered here, their semantic and situation model alignments are parts of the coordination of inner worlds (level 2). Their account does not cover the coordination of meanings (level 3) that is a central part of the semantic processes I discuss in this chapter.

5.3   The Role of Generics

On level 3, when coordinating their meanings of words, the partners may rely on *definitional generics* as a special tool (Lawler, 1973). For example, "whales are mammals" and "a wrench is a tool for fastening nuts" are used to express elements of the meaning of "whale" and "wrench." In contrast, *descriptive generics*, such as "whales are common around Cape Horn" and "wrenches are expensive in this shop" express facts about the world that are not part of the meaning of the words. The basic difference between definitions and descriptions can be expressed in the following way: Descriptive generics presuppose that the meanings are fixed and are common to all participants. They add to the common ground. In contrast, definitional generics provide information about the meanings of the words that are being used. The speaker wants to introduce a new word or to impose a certain meaning constraint on an already available word.[8]

It is interesting to note that the two types of generics behave linguistically in different ways, as pointed out by Lawler (1973):

(5.1a)   Madrigals are polyphonic.

(5.1b)   A madrigal is polyphonic.

(5.2a)   Madrigals are old-fashioned.

(5.2b)   *A madrigal is old-fashioned.

(5.1a) is a definitional generic describing a central part of the meaning of "madrigal." It can be exchanged for the indefinite singular version in (5.1b). In contrast, (5.2a) is a descriptive generic that says something factual about madrigals that is not considered part of the meaning of "madrigal." A test for this is that it cannot be exchanged for the indefinite singular version in (5.2b) (Krifka, 2012a, b; Carlson, 2009).[9]

Thus the two types of generics are used for different types of speech acts, that is, they have different communicative roles. In terms of the three levels of communication, a definitional generic is used as a communicative repair on the highest semantic level, which helps the interlocutors coordinate their use of the terms. On the other hand, a descriptive generic is typically used on second level when the interlocutors need to coordinate their inner worlds: one person fills a gap in the knowledge of another (Krifka, 2012b).

5.4   Coordination as Reaching Fixpoints

Of all affairs, communication is the most wonderful. That things should be able to pass from the plane of external pushing and pulling to that of revealing themselves; and that the fruit of communication should be participation, sharing, is a wonder by the side of which transubstantiation pales.
—John Dewey

The distinction between three levels of communication entails two major types of coordination: one where the participants agree on the meanings but have to coordinate their knowledge about the facts of the world, and one where they have to adjust their understanding of the meaning of a term to be able to return to the lower levels of communication. An analogy is that the first type is like two persons coordinating their positions via a map, while the second is like them coordinating by updating their maps.

The mechanisms relating to fixpoints that I present in this chapter pertain to both types of coordination. Before I present a more technical characterization of fixpoints, I will give some examples of the two types of coordination processes and the fixpoints they result in.

5.4.1   Coordination of Worlds

Coordination can occur in many ways without using words. For a simple example of convergence to a fixpoint as the result of coordination of referents in the world, consider again joint attention achieved via pointing, for example, a child pointing out something to an adult.[10] The goal of the child's pointing is to make the adult react by looking at the desired point in her visual field. The fixpoint is reached when the child sees that the adult's attention is directed at the correct location, and the adult believes her attention is directed to what is being pointed at.

A famous example from Schelling's (1960) analysis of coordination is a situation in which two people wish to meet in New York City on a given day but cannot communicate the time or place at which they should meet. He suggested that beneath the clock at Grand Central Station at noon would be a *focal point* for coordination. Since Schelling's analysis, a number of studies of such focal points have been made. One important factor in this context is that finding a coordinating point depends on advanced intersubjectivity. In the example, each person has to imagine where and when the other thinks would be a good coordination point, and imagine what the other thinks about what she thinks herself would be a good coordination point, and so on. In other words, the persons must figure out what joint beliefs they are likely to have.

When words are available, it is simpler to coordinate. As an illustration of coordination of knowledge about the facts of the world, consider an example involving *definite reference* (H. Clark, 1992, p. 107). An example is a simple communicative act such as explaining to a tourist where to find a restaurant she is looking for: it involves a complex series of further requests and information extensions, as well as corrections, nods, and interjections. Creating such a reference is a coordination problem that is rarely reduced to uttering the right word at the right time. What is required instead is a process of mutual adjustment between speaker and addressee converging on a mutual acceptance that the addressee has understood the speaker's utterance. The process is highly iterative, involving a series of reciprocal reactions and conversational moves usually concluded by assent signals. Such a process has the clear nature of arriving at a fixpoint. Conversational adjustments toward mutual agreement typically resort to both the discrete resources of spoken language and the continuous resources of gesture, intonation, and other bodily signals.

5.4.2   Coordination of Meanings

Aprender a hablar es aprender a traducir.[11]
—Octavio Paz

Learning a language is perhaps the most general way of coordinating meanings of words. For example, when children learn words, they first overgeneralize and then narrow down the meanings.[12] For example, McDonough (2002) studied word comprehension where distractor elements were drawn from the same superordinate category, as well as from different superordinate categories. She found that two-year-olds overgeneralized many words in the animal, vehicle, food, and clothing domains. For example, a rocket was labeled "airplane" and a raccoon "cat." However, overextensions were rare across domain boundaries. When the child learns new prototypes for other concepts within a domain, the child will gradually adjust an early concept to its normal use, since the child's partitioning of the domain will become finer. This process can be modeled using Voronoi tessellations, but I will not spell out the details here.

For another example, consider a semantic negotiation game, in which the interlocutors have an interest in agreeing on a meaning, but each of them wants agreement on different things. This resembles a contractual negotiation. The agents may have a partial overlap in representing a certain requisite concept, but to reach agreement, they need to negotiate a sufficiently common meaning.[13] The process is obviously fallible. One can find many examples of contractual breaches that originate from different meanings associated with agreed contractual terms. In a famous court case, the contractors could not agree on the meaning of the chicken that was to be delivered.

We can nevertheless communicate even if we do not have identical mental representations. For example, in communication between children and adults, children often represent their concepts using fewer domains and domains that have different prominence from those of the adults.

A difference between coordinating worlds and coordinating meanings is that signaling and pointing concern single referents—objects, places, or actions. In contrast, when meanings are coordinated, single referents are not sufficient, but *concepts* become necessary. Words refer to *types* of entities: nouns to types of objects, verbs to types of actions, and adjectives to properties of objects and relations between them (see Gärdenfors, 2000, chap. 5; and part 2 of this book).

Experiments on the emergence of human communication systems often reveal a similar process, whereby mutual understanding emerges as a fixpoint from a joint coordination effort. Some of these experiments use (graphical) sign systems, with the advantage that such systems preserve the continuity of conceptual spaces while verbal systems discretize it—a problem I analyze later on. For example, Galantucci (2005) created a virtual environment (reminiscent of Garrod & Anderson's [1987] setup) consisting of rooms in which players were located without being able to see each other. They had to meet in the same room using a limited number of moves, communicating only through a system of signs on an electronic pad. Sending a sign corresponded to attempting to push the mind of the receiver to locate the right region of the virtual space, while understanding the meaning of the received sign corresponded to mentally locating the correct region. So long as participants did not correctly understand the signals, the communication system remained out of equilibrium, with participants adjusting their reaction to the signs received (and adjusting the signs they sent) until they found a response in which both correctly located the meaning of the sign.

5.4.3   A Signaling Game

La parole est moitié à celui qui parle, moitié à celui qui écoute.[14]
—Michel de Montaigne

The world of computers offers new possibilities to model the language learning process. In addition to the experimental studies presented in the previous section, Jäger and van Rooij (2007) provide a computer simulation that constitutes an example of how fixpoints (or, in their case, Nash equilibriums) can be used to model a meeting of minds. The domain they choose is called the color space, though nothing in the process depends on relations to colors (except that the space used in the simulations is a circular disk). The problem they examine is how a common meaning for "color" terms can develop in a communication game. Their example uses only two players: s (signaler) and r (receiver). Jäger and van Rooij assume that the two players have a common domain C for "color." There is a fixed and finite set of n messages ("words") that the signaler can convey to the receiver.

The goal of the communication game is to maximize the average similarity between the original point and the interpretation of the receiver. The communication game unfolds as follows: Nature chooses some point in the color space, according to some fixed probability distribution. The signaler s knows the choice of nature, but the receiver r does not. Then s is allowed to send one of the messages to r. In response, the receiver r picks a point in the color space. In the game, s and r maximize their rewards if they maximize the similarity (minimize the distance) between nature's choice and r's choice of point.[15]

The signaler can choose a *decomposition* S of C in n subsets, assigning to each subset a unique message. For each color "word" sent, there is a prototypical point in the region corresponding to the word that is the responder's best response on average. In other words, there are n prototype points, corresponding to the typical meaning assigned to each of the n messages from the signaler.

Following the standard definition from game theory, a Nash equilibrium of the game is a pair (S, R), where S is the sender's partitioning (in n subsets) of C, and R is the responder's n-tuple of prototype points of C, such that both are a best response to each other. Jäger and van Rooij (2007) show how to compute the best response functions for each player. The central result of their paper can be restated by saying that if the color space is convex and compact, and the distribution and similarity functions are continuous, then there exists a Nash equilibrium, and it corresponds to a Voronoi tessellation of the color space that is common to s and r (fig. 5.2).[16]

The prototype points have the property that they minimize the average distance to all the points in the region. Note that infinitely many equilibrium solutions exist. Which one is reached depends on the idiosyncrasies

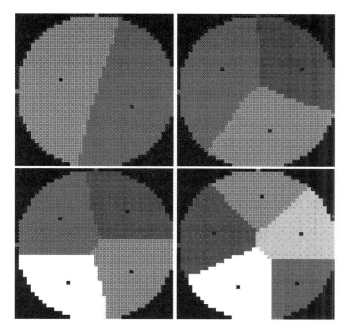

Figure 5.2
Results of Jäger and van Rooij's (2007) simulations of meaning coordination when two, three, four, and five "color words," respectively, were used in the communication. The dots mark the prototype points that are the optimal responses to each word.

of a particular simulation, for example, the sender's first division of the color space.[17]

Jäger and van Rooij's model is interesting also because it provides an illustration of how a discrete system of signs (there are only n signs in their communication game) can give rise to approximations of continuous mappings between agents' mental representations. In their example, signs define an array of locations in the color space, and the "best response function" of s and r continuously maps configurations of such an array of points as responses to decompositions of C, and vice versa.

Apart from the work by Jäger and van Rooij (2007), a variety of simulations and robotic experiments (e.g., Hurford, 1999; Kirby, 1999; Steels, 1999; Kaplan, 2000; Steels & Belpaeme, 2005) have shown that a stable communicative system can emerge as a result of iterated interactions between artificial agents, though no one determines any "rules" for the meanings of the communicative acts. A general finding of the experiments is that the more "speakers" and "hearers" are involved in communication about the same outer world, the stronger is the convergence of the reference of the "words" that are used and the faster the convergence is attained. Still, different "dialects" in the simulated community often emerge. The "mental spaces" that have been used for robots in these simulations have, however, been extremely simplistic and assumed to be identical in structure for all individuals. I return to this topic in chapter 14.

Now, even if meanings are determined by what is in the heads of the users, this does not imply that meaning will necessarily change if the set of users change. People die and new individuals are born into a language community, but the emergent meanings may stay constant even if the speakers are gradually replaced. The fixpoint (language game equilibriums) will make newcomers in a language community adapt to the shared meanings, and in this sense, the meanings will be independent of the people who carry them in their heads.

5.5   Modeling Fixpoints in Communication Games

The basic idea of communication as seeking fixpoints of meaning was expressed a long time ago by Dewey (1929, p. 178):

The characteristic thing about B's understanding of A's movement and sounds is that he responds to the thing from the standpoint of A. He perceives the thing as it may function in A's experience, instead of just ego-centrically. Similarly, A in making the request conceives the thing not only in its direct relationship to himself, but as a thing capable of being grasped and handled by B. He sees the thing as it may function in B's experience. Such is the essence and import of communication, signs and meaning. Something is literally made common in at least two different centres of behavior.

In this section, my goal is to express this idea more strictly using some fundamental mathematical notions and results.

5.5.1   The Semantic Reaction Function

There are some important issues relating to the semantic use of fixpoints. One is simply the existence of such fixpoints, and what features of agents' conceptual spaces may favor them. Another concerns how to reconcile the continuous nature of mental spaces with the inherent discreteness of verbal language. A third is how the values of communication shape semantic interaction, in particular how fixpoints are achieved. I will deal in turn with each one of these issues.

The "meeting of minds" semantics result from an interactive, social process of meaning construction and evaluation. With the examples of coordinating worlds and coordinating meanings as a background, I now turn to a more precise way of describing fixpoints. Mathematically a *fixpoint* is a value x that a function f maps x onto itself: $f(x) = x$.[18] The next task is to describe the relevant function f.

Before this concept can be applied, I must explain what kind of semantic function is involved when minds meet. In communicative acts, two functions are involved, as hinted at in the quotation from Dewey. One is the *expression function* that maps the speakers' mental representations (represented as points in conceptual spaces) onto expressions (words or gestures). The second is the *interpretation function* that maps expressions onto the hearer's conceptual space. The composition of these functions constitutes the *semantic reaction function*.

As an example, consider what happens in Jäger and van Rooij's (2007) game. When a color c is selected by "nature," the expression function describes how the signaler selects a signal expressing the color. In the equilibrium situation, the signaler will associate each message with a prototypical color. He selects a message m, the prototype of which is closest to the color c. When the receiver r hears m, it is mapped by the interpretation function onto her color space. Again, in the equilibrium situation, after the learning phase, the interpretation function maps m onto the prototype of a color region for both the speaker and the receiver. Conversely, when the receiver sends back the message m, it will be mapped onto the signaler's color prototype. In the equilibrium situation, the semantic interpretation function will consequently map each color prototype onto itself. Hence the prototypes are fixpoints of the semantic reaction function (fig. 5.3).

Clark (1992, pp. 110–112) presents a series of examples of how definite reference is made common in conversations. An example is the following (adapted from H. Clark, 1992, p. 111):

B: How long will you be here?
A: Not too long. Just until Monday.
B: You mean a week from tomorrow?
A: Yeah.

In this example, the initial reference of "Monday" is ambiguous, and B's follow-up question makes the day a common reference for both speakers. Clark's examples provide more complex illustrations of the nature of semantic reaction functions and the reaching of fixpoints. Cases in which

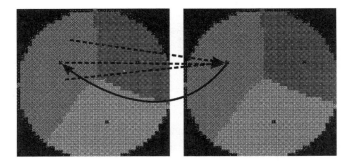

Figure 5.3
Prototypes as fixpoints of the expression-interpretation function.

the conversation participants can jump directly to a commonly agreed reference point are relatively rare. Reference is usually achieved through a sequence of steps. The speaker must choose one out of many possible utterances—typically a noun phrase—to initiate the process. The initial move may be mapped by the addressee into a state of mind inadequate to select the reference, in which case a sequence of adjustments will be triggered, requiring moves such as expanded descriptions, elaborations on or repairs to the original phrase, attempts to verify that the addressee has understood the reference, and so on, until signals of mutual agreement are produced.

Communicative acts are, as Clark (1996, p. 23) emphasizes, a joint activity that involves both the speaker's meaning and the addressee's understanding. Hence, in contrast to traditional semantic theories, both an expression and an interpretation function are required to model the process of a meeting of minds. Clark's (1992, 1996) analyses of common ground and uptake show that the *cycles* of signals and responses, rather than single utterances, are constitutive elements in communication. The role of such cycles in creating meaning has been underestimated in most semantic theories.

5.5.2 The Fixpoint Theorem
I next show how some topological and geometric properties of mental representations make meetings of minds possible because they generate communicative fixpoints in a natural way. Following the proposal in chapter 2, I assume that concepts can be represented as convex regions of conceptual spaces. However, I assume neither that the spaces of the communicating individuals are identical nor that they partition the spaces in

the same way. For technical reasons, the spaces will also be assumed to be compact.

A basic tenet of cognitive semantics is that language preserves the spatial structure of concepts. One way to express this is that language preserves the *nearness relations* among points in conceptual spaces. It is important to remember that the conceptual spaces of the interlocutors are based on similarity, which provides the spaces with their metric structure: in conceptual spaces, "near to" means "similar to." In topology, a nearness-preserving function is nothing other than a *continuous* function. In other words, assuming that language is able to preserve nearness relations implies assuming that language is able to establish a continuous mapping between the mental spaces of different individuals. The assumption basically says that natural language must have enough plasticity to map near points in one conceptual space onto near points in another conceptual space.

In Warglien and Gärdenfors (2013), the following result is shown to be a direct consequence of Brouwer's (1910) famous fixpoint theorem:

Semantic fixpoint theorem: Every semantic reaction function that is a continuous mapping of a convex compact set onto itself has at least one fixpoint.

In simpler words, no matter how individual conceptual spaces are structured and how individual representations of concepts are partitioned, as long as such representations are convex and compact and the communicative device plastic enough to preserve the spatial structure of concepts, there will always be at least one point representing a meeting of minds. I present the mathematics behind the theorem in greater detail in the appendix.

Seeing semantics as a meeting of minds involves shifting from a conventional emphasis on the way we share (the same) concepts to an emphasis on the way the shapes of our conceptual structures make it possible to find a point of convergence. Just as wheels are round because they make transportation efficient, convex conceptual shapes should be selected because they make communication smooth and memorization efficient. Communication works well so long as it preserves the structure of concepts.

The mapping from the conceptual space of one individual to that of another need not go via a discrete language, either verbal or signed. Other kinds of communication can be used, such as gestures, mimicry, and other visual tools. Using these means, people can construct continuous functions between their mental spaces. The specific shape of such continuous functions will depend on a variety of pragmatic factors.

After this mathematical detour, my central claim should be clear: whenever it is important to reach a meeting of minds, representing concepts as convex regions provides the background for language to deploy most of its power. I am not claiming that convex representations are faithful representations of the world—only that, since they are effective, one should find them quite widespread.

An advantage of the fixpoint approach is that it allows one to consider a wide variety of types of communicative interaction, corresponding to different game types. For example, one might want to distinguish between coordination and negotiation games. In coordination games, such as the Jäger and van Rooij (2007) color game, the participants share common interests. In negotiation games, the participants have an interest in reaching agreement, but they have diverging interests on which agreement to reach.

The fixpoint theorem guarantees that minds can meet, but it does not reveal anything about *how* the semantic reaction function emerges as a result of communicative interactions. The next section deals with this topic.

5.6   Determining the Semantic Reaction Function

Communication always occurs for some reason. I have postponed until now the issue of how motivation and the stakes of communication influence the semantic reaction function. Given what I have presented, it might seem that the semantics I propose are purely mentalistic, construed from the conceptual spaces of the interlocutors, without any grounding in the external world. Reality enters, however, via the *payoffs* of communication. If meanings are not properly aligned with reality, communication will be unsuccessful and may incur costs. Reality is the reason why not all our wishes come true—or, as Philip K. Dick expresses it, "Reality is that which, when you stop believing in it, doesn't go away."

Reality enters communication, for example, when we use indexicals. A paradigmatic case is pointing in combination with saying *this* or *that*. As I discussed in the introduction to this chapter, pointing is a way of coordinating our visual spaces, via reference to an external world where a joint reference is a fixpoint. Hence pointing and other means of direct reference are central ways of *grounding* words in reality.

In Jäger and van Rooij's (2007) game, explicit payoffs were introduced based on the success of the communication; the payoffs generated the best response function that determined the fixpoints. Once the pragmatic factors that determine communication payoff are considered, it becomes natural to view communication as games where speech acts become moves that modify the conversational playground.

Schelling's (1960) coordination game has often been used to introduce simple forms of semantic equilibriums (fixpoints). In such games, all equilibriums have the same payoff; a joint selection must be made by resorting to external factors, such as conventions or perceptual salience (see also Lewis, 1969). Of course, reality may enter such games by making the payoffs of some equilibriums superior. To make a Schelling-style example: Imagine that you have agreed to a meeting in a very large square. There are infinitely many equilibriums corresponding to the different locations in the square. However, it happens to be raining, and only one spot in the square is protected from rain. The obvious equilibrium to be selected is the one with the payoff of being protected from the rain. In this way, reality payoffs can be used to select equilibriums in communication games. A related example once more involves definite reference. While doing some electric repair work, Victoria asks Oscar to bring her the screwdriver. There are two identical screwdrivers in the room; one close to Victoria and Oscar, the other much more distant. Obvious reasons of payoff dominance suggest the closer screwdriver as the referent.[19]

In other cases, reality may disrupt meetings of mind. Small children sometimes confuse height attributes with age. A mother's description of an adult as older than an adolescent may generate, in the mind of her child, the conviction that the adult is taller than the adolescent. The child's signs of assent may convince the mother of mutual understanding—but when the tall adolescent shows up, mutual understanding may collapse.

The semantic reaction function need not be the result of an agent's explicit representation of the mental space of her interlocutor. As Pickering and Garrod (2004) have argued convincingly, many dialogic moves are primed automatically by the conversational context, resulting in the convergence of interlocutors with a remarkable cognitive efficiency. An example of such automatic reactions is the tendency of listeners to complete the utterances of a speaker before she has finished producing them.

If the mental spaces of the interlocutors diverge widely, for example, by the domains that are available to them, more radical communication methods must be applied to achieve a meeting of minds. This is where metaphors become powerful tools. By using a metaphor that exploits a shared domain, the speaker can convey information about another domain that has no or only vague correspondence in the semantic domains of the listeners.[20] For example, if one wants to express an emotional state that goes beyond the experiences of one's listeners, a metaphorical description is often the only available resource.

Finally, a meeting of minds need not mean an identity of representations between agents. For example, the color space of a partially color-blind person may have a different geometric structure (or even fewer dimensions) than that of a person who has normal color vision. It might even be impossible to conceive a perfect alignment of representations among conversation participants (Parikh, 2010), though some alignment is part of the convergence process in most conversations. (Pickering & Garrod, 2004). A strong aspect of the fixpoint approach is that it can encompass communicative agreement between participants even when their conceptual spaces differ. What makes communication possible is the capacity to establish similarity-preserving mappings between the conceptual spaces of the participants and to approach mutual fixpoints.

5.7   Expanding the Common Ground

The composition of meaning usually takes place in multiple steps in extended communication structures such as conversations. A full account of the composition of meaning should therefore consider the sequential process through which meanings are determined.

An *assertion* is a typical move in a communication game that proposes to enlarge the common ground (H. Clark, 1992; Warglien, 2001). Stalnaker (1978, p. 323) writes: "The essential effect of an assertion is to change the presuppositions of the participants in the conversation by adding the content of what is asserted to what is presupposed. This effect is avoided only if the assertion is rejected." If the listener's countermove is accepting the assertion, then both parties expand their common ground. If, on the other hand, the listener rejects the assertion, another move will need to be attempted (for examples, see H. Clark, 1992).[21]

The common ground functions as a source background knowledge that can be used by the interlocutor to make inferences and predictions. Grice's (1975) conversational implicatures and Traugott and Dasher's (2002) invited inferences are examples of analyses that pick up this role of the common ground. In Gärdenfors (2000) I also account for how the concept structure generates different forms of nonmonotonic reasoning based on expectations.

Another account is that of *discourse representation theory* (Kamp, 1981; Heim, 1982). This theory presents so-called discourse representation structures that build up as the discourse unfolds. Every new sentence prompts additions to that representation. In particular, the theory includes a formalism that has been developed to handle anaphoric pronouns (which

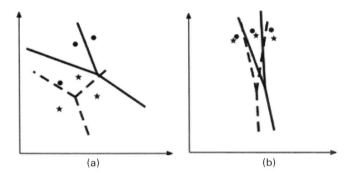

Figure 5.4
Agreement by prominence adjustment.

depend for their interpretation on earlier expressions in the discourse).[22] The meanings of pronouns and indexicals are determined by the fast convergence process. Unlike other words, their meanings change as soon as the context of the discourse changes. This point has been overlooked in most semantic theories.

If the communication game constantly expands the shared mental spaces of the participants, convergence is intuitively unavoidable. The typical rules of conversation, as set down by Grice (1975), are communicational institutions that help maintain the expansions of common ground and thus ensure convergence.[23]

The context of the discourse also influences the relative weight of the domains. A simple visual example will illustrate this point. Two agents with different sets of prototypes (dots and stars in fig. 5.4a) may generate different decompositions of a two-dimensional space (solid and dotted lines), such that their categorization of entities in that space will disagree most of the time. Nevertheless, as conversation proceeds, they may alter their representations, for example, by assigning a greater prominence to the x-axis. Changes in prominence will result in changes of axis weights, altering the distance between points (Gärdenfors, 2000). By adjusting the weight of the two dimensions, the agents will be able to align their categorization of points in the space, so that they will come to agree most of the time (fig. 5.4b).

5.8   Meetings of Minds Do Not Leave the Minds

How can we test whether minds really meet? In a case where the communication concerns something that is present in the environment, reality constrains the payoffs of the communication, as we saw earlier. When the communication is not about the present or is about fictional objects, it becomes difficult to check the meeting of minds. Although a speaker gets feedback, signaling acceptance of a speech act from the point of view of the listener, and the listener believes that she has understood the speech act, we have no guarantee that their minds really meet. A common belief of understanding may exist, but there is no standard external to the minds of the interlocutors to judge whether they "really" mean the same thing. In this sense, a meeting of minds is a construction of the minds. The only thing that can be achieved is that the interlocutors mutually believe that they understand each other. The interaction has to rely on this assumption, though there may be no way of externally confirming the meeting of minds. Thus the theory does not contain any notion of reference in the traditional philosophical sense. Instead my semantic theory builds on how the individuals construe the world in their minds and how they coordinate their construals.

An illustration of this is a well-known philosophical problem concerning *intentional identity*, by which is meant that several people "have attitudes with a common focus, whether or not there actually is something at that focus" (Geach, 1967, p. 627). A variation of Geach's classical example is the following:

(5.1)   Hob says: "A witch blighted Bob's mare."

(5.2)   Nob says: "She has killed Cob's sow too."

The problem is how Hob and Nob can know whether they are speaking about the *same witch*. Standard techniques within logic fail to give an acceptable answer. With the exception of Edelberg (2006), few proposed solutions have taken into account the communicative aspects of intentional identities. Jacot (2012) uses the meeting-of-minds semantics to analyze the problem. A first observation is that, just as for any referential expression, Hob's mental representation of the witch need not be identical to Nob's representation of the witch; it is sufficient that they share enough properties so that their communication reaches a fixpoint. Since witches do not exist in the real world, the solution must rely entirely on the representations in Hob's and Nob's conceptual spaces.

The use of the expression "a witch" introduces a reference to an individual, but the reference is underdetermined. The most accessible meaning for a listener is the *prototype* for a witch. Jacot (2012) argues that if it is assumed that Nob identifies the witch with an individual who has all the prototypical properties of a witch (in his mental spaces), then this is sufficient to establish a fixpoint that functions as a reference for the witch. To be more precise, when Hob uses the expression "a witch," it is mapped onto the point in Nob's conceptual space that represents a prototypical witch. When Nob then uses "she" in his response, he indicates that his prototype has the property "female." As long as Hob's prototype also has this property, Nob's use of a pronoun indicates an "uptake" (H. Clark, 1996) of Hob's introduction of a new referent to the common ground, and Hob interprets "she" as referring to the same individual that he referred to with "a witch" (which among other things entails that he too considers the witch to be female). If Hob also adds to the common ground that the witch killed Cob's sow, a fixpoint is reached in communication.

However, the fixpoint that is established via the prototype is fragile. If, in a continuing discussion about the witch, Hob asserts that she has a property (e.g., not being able to cast spells or not being present in the vicinity when Cob's sow was killed) that is incompatible with the properties of Nob's prototypical witch, then the fixpoint breaks down, and they both have to retract the uptake of the referent in their common ground. In this case there is no longer any intentional identity.

Despite the problems of intentional identity, there is still a way to test to which extent the minds of the communicators meet: If their minds are coordinated, their *inferences* should also be coordinated. In particular, the communicators should be able to predict the beliefs of the other partner (see Andersson, Holsanova, & Holmqvist, submitted). For example, suppose that A, who lives in Sweden, and B, who lives in the United States, have agreed to meet in Cambridge next Tuesday, then A may draw the conclusion that B will fly to England within a few days, and B may draw the conclusion that A will soon fly to Boston. If these inferences are made explicit, their lack of coordination will be revealed, and the meeting of minds may be repaired.

5.9 Conclusion

In this chapter, I have described the semantic interpretation function on an abstract level. In verbal communication, messages are mediated by words. To flesh out the structure of the interpretation function, we must therefore consider the contribution of different types of words. The second part of the book will be devoted to a reconsideration of lexical semantics on the basis of conceptual spaces as a common modeling tool and from the general perspective of semantics as a meeting of minds.

I conclude the chapter by outlining how the theory of semantics as meetings of minds handles the six criteria of section 1.2. Regarding the ontological criterion, this theory is still a cognitivist theory, since it builds on individual mental representations of meanings. However, since it also takes into account the interplay between different communicators, it can be called a *sociocognitive* theory.[24]

To answer the semantic question is to describe the relation between linguistic expressions and mental constructions. Unlike the versions of cognitive linguistics presented earlier, the relevant mental constructions are emergent meaning equilibriums (fixpoints) in the community of users. As I show later in the book, the relation between words and meanings to a large extent also depends on the communicative *context*. This feature puts my theory in stark contrast to most other theories that see semantics as a more or less fixed mapping between words and meanings.[25] Consequently the theory does not build on a sharp distinction between semantics and pragmatics.

The epistemological criterion is treated in terms of associative learning: a person grasps the meaning of an expression by relating it to a cognitive structure. However, to give some substance to the answer, more must be said about how the connection is established. Here de Saussure's (1966) statement that "the two elements involved in the linguistic sign are both psychological and are connected in the brain by an associative link" provides the central clue: a theory of learning based on *associations* will construe the coupling between a linguistic expression and its cognitive meaning as just a special case of learning in general. In chapters 3 and 4, I have shown how this learning depends on meaning being structured in domains. However, the lion's share of the learning of the semantic mapping comes from interactions with other individuals. This also means that the social criterion is built in as a central part of the semantics.

Finally, this kind of semantics handles perception and action via the anchoring in many domains of the conceptual spaces. This anchoring is more general and systematic than what is found in most versions of cognitive semantics. In chapter 8, I develop an account of how actions derive from force spaces as a basis for verb semantics, and in chapter 13, I show how compositions of meanings can be generated in a systematic way.

The upshot is that the semantic theory based on meetings of minds satisfies all the six criteria presented in chapter 1.

II   LEXICAL SEMANTICS

Ich fürchte mich so vor der Menschen Wort.
Sie sprechen alles so deutlich aus:
Und dieses heißt Hund, und jenes heißt Haus,
Und hier ist Beginn und das Ende ist dort.

[The words of humans fill me with fear
They name everything with articulate sound:
So this is called house and that is called hound,
And the end's over there and the start's over here.]

—Rainer Maria Rilke (translation by Walter A. Aue)

6   Object Categories and the Semantics of Nouns

6.1   Cognitive and Communicative Foundations for Word Classes

Why are there words at all? Why do we not communicate with wordless songs as the wolves and whales do? The short answer is that we have words because we need to communicate about *categories*—above all, object, action, and event categories.

Most researchers within semantics look at the meaning of words from a linguistic perspective. From this perspective, it is difficult to free oneself of syntactic concepts. For example, the "arguments" of verbs show up in most semantic analyses (e.g., Levin & Rappaport Hovav, 2005). The notion of argument derives, however, from syntax. Among other things, this leads to the distinction between transitive and intransitive verbs. However, this distinction does not correspond to any clear-cut semantic distinction. Similarly, it is said that verbs and adjectives are used in a "predicative" manner. The notion of predicative derives from theories in philosophy and linguistics that aim at grounding semantics in predicate logic. In my opinion, this is an artificial construction that does not have a cognitive grounding. In contrast, my ambition is to develop semantic models that are constructed from general cognitive mechanisms. The semantic theory of this book is supposed to be syntax free. In other words, the semantic notions should not depend on any grammatical categories. This does not mean that I deny that syntax contributes to meaning (Langacker, 2008, pp. 3–4). I only claim that lexical semantics can be treated independently from syntax.

One of the most fundamental concepts of linguistics is that of *word classes*. In all languages, words can be grouped in distinct classes with different semantic and syntactic functions.[1] Some variations exist between languages, as I discuss later. In English the words have traditionally been classified into eight classes: nouns, pronouns, adjectives, verbs, adverbs, prepositions, conjunctions, and interjections.[2]

Within linguistics, a word class is defined in grammatical terms as a set of words that exhibit the same syntactic properties, especially concerning inflections and distribution in sentences. I do not believe in a universal definition of word classes (cf. Croft, 2001, chap. 3). Syntactic structure, including the division into word classes, is language specific. However, one can identify prototypical structures among words that can be used in classifications. Croft (2001, p. 63) writes: "Noun, verb, and adjective are not categories of particular languages. . . . But noun, verb, and adjective are language universal—that is, there are typological prototypes . . . which should be called noun, verb, and adjective."

Communication is produced for a purpose in a particular context. That major word classes such as verbs, nouns, and adjectives can be identified in almost all languages suggests that universal patterns in human cognition make the division into these classes particularly useful for communication (Dixon, 2004). The structure of communication is subject to the same cognitive constraints as thinking and problem solving in general. Therefore it is reasonable that the structure of language, at least to some extent, is determined by such general cognitive principles. In particular, I assume that the structure of language is governed by the same principles of processing efficiency of representation as are other cognitive processes.

When word classes are taught at an introductory level in school, semantic criteria are used, for example, that nouns stand for things and verbs describe actions, but these are seldom presented in a systematic and rigorous way. One aim in the second part of this book is to show that word classes can be given a (syntax-free) semantic analysis based on cognitive and communicative constraints.[3]

In this and the following chapters, I specify what I take to be the main communicative functions of the major word classes. Since I have argued that the pragmatically (and evolutionarily) most important role of communication is to coordinate actions (level 1), I focus on situations of this type. As a paradigmatic case, I take communication about a nonpresent referent and how coordination about such a referent is involved in planning for joint actions.

As I have already indicated, the distinction between properties and concepts is useful for analyzing the cognitive role of different word classes. In brief, I propose that the main semantic difference between adjectives and nouns is that adjectives like "red," "tall," and "round" typically refer to a *single* domain and thus represent properties, while nouns like "dog," "apple," and "city" contain information about *several* domains and thus represent object categories. This characterization is, however, just a rule of thumb, and I develop it further in this and the following chapter.

6.2   Object Categories

The Things
The things have weight,
mass, volume, size,
time, shape, color,
position, texture, duration,
density, smell, value,
consistency, depth,
boundaries, temperature,
function, appearance, price,
fate, age, significance.
The things have no peace.
—Arnaldo Antunes

In this chapter the concepts in focus are *object categories*, that is, concepts that are used to classify objects.[4] In addition to properties from the domains that have been illustrated, object categories often depend on *meronomic* relations, that is, information about which parts entities have and how the parts are related.[5] For the shape domain, which is central in the classification of many physical objects, the meronomic relations can be seen as included in that domain. However, meronomic relations do not exist only in the shape domain. For example, possessive relations, such as "a farmer owning a plow," are also construed as meronomies; the farmer's assets have parts and subparts. Another example where meronomic relations are found is social structures. For example, *clans* have families as parts, which in turn have members.

As a paradigm example of an object category, consider "dog." When as children we encounter dogs, among the domains that we learn about are presumably shape, size, sound, smell, color, and texture. We also learn about the *material* of different categories, for example, that dogs, like other animals, consist of flesh and bones. Thus the material domain is important for categorization, in particular for categories of natural objects.[6] In parallel, we learn about further domains of dogs, such as their typical behavior (the action domain), biological characteristics, and their roles in human society. We also learn about the parts of a dog and how they are related. Dogs vary a lot in their looks, so most of their meronomic relations are shared with all four-legged mammals, but some parts, typically tail, ears, and nose, have characteristics that are more or less unique to dogs (or at least canids). Again, knowledge about the meronomic aspects of a category develops over time. For example, small children confuse dogs with calves, since they do not notice the differences in the structure of the nose or muzzle part of the animal and the differences in their tails.

In this section, I discuss the domain aspects of object categories, and in the following section I take up the meronomic aspects. The first problem when representing a category is deciding which are the relevant domains. I should note that I do not require that a concept be associated with a closed set of domains. On the contrary, the domains associated with a category may be expanded as one learns about further aspects of a category.[7] The addition of new dimensions and domains is often connected with new forms of actions and interactions that require attention to previously unnoticed aspects of categories. For example, when a child learns about magnetism, a new dimension that generates forces, he or she often becomes interested in which kinds of objects are attracted by magnets.

Furthermore, not all domains are of equal importance for an object category. When several domains are involved in a representation, there must exist some principle for how the different domains are to be weighed together. To model this, I assume that the category representation also contains information about the *prominence* of the different domains.[8]

Langacker (1987, p. 165) calls a domain that is highly ranked a *primary* domain.[9] He illustrates the notion by the difference in meaning between "roe" and "caviar." The two categories refer to the same entities (masses of fish egg). For "roe," however, it is the zoological domain of fish that is prominent, while for "caviar," the domain of food is the most prominent.[10] Note that the relative prominence of the domains depends on the context in which the category is used. For example, if you are eating an apple, its taste will be more prominent than if you are using an apple as a ball when playing with an infant, which would make the shape domain particularly prominent.

Sometimes the same word is used with two different foci. A standard example is "book," which can mean both the physical object ("the book has hard covers") and the information content of the text in it ("the book is difficult to understand"). In my approach, this is again a question of which domains are prominent. When "book" is used about a physical object, the object category space is in focus. However, when "book" refers to its information content, the physical properties become more or less irrelevant. The information content belongs to an abstract, nonphysical domain.[11] As a matter of fact, two copies of the same book are *identified* as the same under this interpretation, since they have identical information content. In this way, the domain of information content can be detached from the physical domains, and we can talk about book in an abstract information space. Similarly, "dollar" can refer to both the physical object and its value in the monetary domain. Again, two dollar bills are seen as identical in terms of value.

The notion of prominence can be used to clarify the much-debated distinction between lexical and encyclopedic components of meaning (e.g., Langacker, 1987, p. 154; 2008, pp. 38–39). The idea is simply that the most prominent domains belong to the lexical part of meaning, and the less important to the encyclopedic part. Since prominence of dimensions comes in grades, a consequence is that no sharp boundary separates the lexical and the encyclopedic components of meaning. I think this is as it should be. For example, is it part of the lexical meaning of "cat" that a cat purrs? For me this is a borderline case.

The distinction between definitional and descriptive generics from section 5.3 can be used as a guiding test of which domains are considered to be included in the semantics of a category: if a generic *an X is Y* is considered definitional, then the domain of Y is part of the meaning for the category X. However, this test does not have a strong independent value.

The next problem is to determine the geometric (or topological) structure of the domains. The color domain can be represented by hue, intensity, and brightness, and the size domain by the height and length dimensions. Other domains are trickier: it is difficult to be precise about the topological structure of "sound space," although pitch and time are fundamental dimensions that can then be used to construct other configurational domains in music, for example, rhythm and harmony spaces. I present some ideas about how "shape space" should be modeled in the following section. Textures could possibly be modeled using dimensions from fractal theory (see, e.g., Pentland, 1986). Nevertheless it would be cumbersome to give psychological support for the geometry of many of the domains. And there are social and abstract domains for which the geometric or topological structures have, so far, not been investigated.

Object categories are not just bundles of properties. The proposed representation for a category also includes an account of the *correlations* between the regions from different domains that are associated with the category. In the "dog" example, a strong (negative) correlation exists between the size of the dog and the pitch of its barking sound.[12]

6.3   Meronomic Relations

Meronomy is a semantic relation between an entity and its parts, for example, the relation between a horse and its four legs. Meronomic relations apply both to individual instances (a particular horse) and to general categories (it is part of the meaning of "horse" that it has four legs). I argued in section 2.6 that meronomies form the basis for metonymies in language. Still, parts are not always necessary for membership in an object category: a horse with three legs is still a horse. For many object categories, meronomic relations are important components, and they should therefore be part of the semantics of such categories. To give a slightly artificial example, when you are reading, you identify the letters of the alphabet almost exclusively by their meronomic relations—fonts can be widely different as long as the parts of the letters have the right relations.

A strong argument for the importance of meronomic relations in concept formation comes from Tversky and Hemenway (1984). They show that part terms occur frequently when subjects describe categories at the basic level, but are rare on superordinate levels. Basic level objects are often distinguished from each other by their parts. Furthermore, subordinate categories share parts with the basic level but differ from one another on other domains.

There exist several analyses of the formal properties of part-whole relations (e.g., Simons, 1991; Varzi, 2003), but I will not present the details here (cf. Fiorini, Gärdenfors, & Abel, submitted).[13] Moltmann (1998) argues that part-whole relations are not sufficient: not any bundle of things forms a whole; there must be some *integrity* of the parts too. However, she does not specify how this integrity is constituted.

For categories referring to physical objects, part-whole relations are generally configurational structures determined from the *shape* domain. It is well known that shapes of objects are crucial for early concept learning. For example, Landau, Smith, and Jones (1998, p. 21) write: "Our view holds that, however the meanings underlying object names are ultimately characterized, shape similarity constitutes a critical bootstrapping mechanism operating to initiate learning of object names in young children by allowing them to identify category members in the absence of dense knowledge about the category."

In other words, the shape domain is highly prominent for children. There are many attempts in the literature to model the shapes of objects (e.g., Marr & Nishihara, 1978; Pentland, 1986; Biederman, 1987; Zhu & Yuille, 1996; Edelman, 1999; Chella, Frixione, & Gaglio, 2001; Petitot, 2011). I will outline here two of these approaches that focus on biological

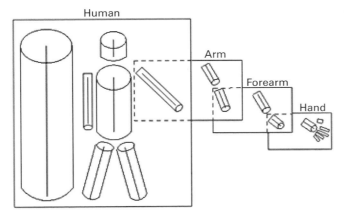

Figure 6.1
Parts hierarchy model of Marr and Nishihara (1978). The arm is decomposed into finer parts along the chain of part structures. Reprinted from D. Marr & K. H. Nishihara. (1978). Representation and recognition of the spatial organization of three-dimensional shapes, *Proceedings of the Royal Society in London, B 200,* 269–294, by permission of the Royal Society.

forms to show how they can be interpreted within a conceptual space approach.

6.3.1   Marr and Nishihara's Model

In object perception, it is well known that the length and the directions of the parts of objects are often sufficient to categorize an object. A classical example is Marr and Nishihara's (1978) scheme for describing biological forms that uses cylinderlike modeling primitives, as illustrated in figure 6.1. The cylinders are combined in a hierarchical manner, with the torso on the first level, and the head and legs (arms) on the second. These parts can then be broken up into subparts, and so on. First, note that each cylinder can be described by two coordinates (length and width). This means that the hierarchy of cylinders can be modeled as a tree-structured vector of pairs of numbers together with values for the connection points of the subparts to the parts.

To achieve further detail in the representation, the cylinders can then be combined by determining the typical *angle* (or range of angles) between the dominating cylinder and the added one (this is not part of Marr and Nishihara's model). The details of the representation are not important in the present context, but it is worth noting that an object can be described by a comparatively small number of coordinates based on lengths and angles. Thus the object can be represented as a vector in a high-dimensional shape domain constituted of values in spatial dimensions and dimensions representing the angles between cylinders.

The vector coordinates for the cylinders and their connecting points will generate a multidimensional space of shapes. Each particular cylinder shape will then correspond to a point in this space. It is possible to introduce a metric on the shape space, for example, by using a weighted sum over all the dimensions on the different levels. A point in this structure space represents a combination of shape forms and displacements for a given decomposition level, denoting a particular limb configuration.

In the multidimensional shape space, one can then look at various regions, perhaps generated from prototypical animal shapes, and investigate their properties. The *metric proportions* between parts play an important part in defining the prototypes of the regions. For example, the *ape* and the *human* shapes have the same parts in basically the same relations, but the proportions between the lengths of the arms and the legs are different for the two categories (see fig. 8 in Marr & Nishihara, 1978). Since Marr and Nishihara's simple cylinder figures are enough to enable us to identify a large class of animal shapes, it seems plausible that much of the content of our concepts for animals comes from their shapes. For instance, a point in the *human* region represents a configuration of the body parts. Intuitively, the *ape* region of shape space is much more similar to the *human* than to the *horse* region (see also Landau & Jackendoff, 1993; Aisbett & Gibbon, 1994).

6.3.2   Zhu and Yuille's Model

Marr and Nishihara's (1978) model shows that the spatial relations between the parts involved in a category can be analyzed as a configurational domain emerging from the domain of physical space in analogy with how I analyzed *rectangle* in section 2.5.4. However, there is more to meronomy: objects are conceived of as parts that *hinge* together; that is, there are cohesive *forces* between the parts. These forces can be seen as an explication of the integrity of an object that Moltmann (1998) calls for. In other words, the shape domain may not be sufficient to describe the meronomic information about a category; the force domain is also required. To see the point, note that shadows have parts, but the parts have no cohesion.

A model that takes the integrity of the object into consideration is presented by Zhu and Yuille (1996). They focus on two-dimensional shapes, but in principle their model can be extended to three dimensions. The shape models consist of two types: *worms*, which are rectangles with joint

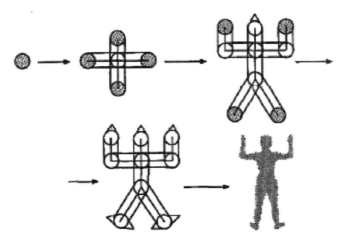

Figure 6.2
Zhu and Yuille's hierarchical model of a human shape. Reprinted from Zhu and Yuille (1996) with permission from Springer Science and Business Media.

circles attached at the end, and *circles* with angles that represent hinges and their flexibility. Like a cylinder, a worm can be described by the two coordinates length and width. A hinge can be described by the radius and the angular values. These primitives can be modified by deformation vectors, but I will not consider these here.

The primitive parts are then combined according to a *shape grammar* that generates a hierarchical structure, for example, the main parts of a human (fig. 6.2). As in Marr and Nishihara's model, this structure can be represented by a multidimensional vector.

Zhu and Yuille (1996, pp. 203–204) present an explicit measure of similarity for comparing different shapes or for comparing a shape to a prototypical model. An interesting feature is that their similarity measure works even if the parts of two objects cannot be matched one by one. For example, a hand with a missing finger is still judged to have a high similarity with a typical hand, and a dog with three legs is still seen as a dog.

These outlines of Marr and Nishihara's (1978) and Zhu and Yuille's (1996) models of meronomic shape relations show that, in principle, the relations can be represented in conceptual spaces, albeit high-dimensional spaces that are more complicated than for other perceptual properties. However, much empirical and mathematical work remains to turn these sketches into practical working models (see Fiorini et al., submitted).

For some objects, different attentional focus on parts of an object can lead to linguistic ambiguities.[14]

(6.1)   He opened the window and went through it.

(6.2)   He smashed the window and went through it.

(6.3)   *He lifted out the window and went through it.

In these examples, "window" is used in two senses: (i) the compound of the pane(s) with its sash, and (ii) the compound of the pane(s) with its sash and the frame in which the compound is hinged. In (6.1) "window" has meaning (ii), and in (6.2) "window" has meaning (i). In (6.3) "window" has meaning (i), and "went through it" normally only makes sense when "it" refers to meaning (ii) This double meaning explains why (6.3) is odd.[15] Similarly, in "He knocked on the door," "door" means the panel, while in "He walked through the door," "door" (normally) means the panel together with its frame and thus "through" refers to the frame.

6.4   Object Categories and Their Relations

6.4.1   Definition of Object Category

Putting together the material from the two previous sections, I now propose the following definition:

An *object category* is determined by

(i) a set of relevant domains (may be expanded over time)
(ii) a set of convex regions in these domains (in some cases, the region may be the entire domain)
(iii) prominence weights of the domains (dependent on context)[16]
(iv) information about how the regions in different domains are correlated
(v) information about meronomic (part-whole) relations.

I analyze criteria (ii)–(iv) in greater detail in Gärdenfors (2000). As we will see later, the prominence of a domain to a large extent depends on the context and on the attentional focus. If meronomic relations are analyzed as configurational domains, condition (v) will be a special case of (ii). However, even if meronomic information is represented as a domain, typically based on the shape domain, its divergent structure justifies that it be treated in a special way. The relative importance of meronomic information may vary between categories. Some categories, such as *hand* and *bicycle*, are to a larger extent determined by meronomic relations, while others such as *apple* and *ball* are more determined by information about regions of other domains.[17]

The regions of the domains of an object category may have vague borders. Furthermore, points in the regions are more or less central. This means that the definition of object categories leads to prototype effects (Rosch, 1978). Generics that are stated about categories, for example, "Ponies have tails" and "Birds build nests," refer to prototypical instances of the category. However, they do not always express necessary conditions: a pony without a tail is still a pony, and some bird species do not build nests.

Condition (iv) introduces correlations between domains as a factor of an object category. An argument for this condition comes from work by Billman (1983) and Billman and Knutson (1996), which indicates that humans are quite good at detecting correlations that cluster *several dimensions*. A plausible explanation of this phenomenon is that our perceptions of natural objects show correlations along multiple dimensions, and as a result of natural selection, we have developed a competence to detect such clustered correlations. In line with this, Holland, Holyoak, Nisbett, and Thagard (1986, pp. 183–184) formulate the hypothesis that the *basic level* categories of prototype theory (Rosch, 1975, 1978) are characterized by distinctive clusters of correlated properties. As explained in Gärdenfors (2000, sec. 6.6), such correlations make it possible to draw *inferences* about categories and their relations.[18]

6.4.2 Hierarchies of Object Categories

The set of domains associated with an object category and its set of meronomic relations is also involved in the *hierarchy* of categories that is reflected in cognition and language. Given my definition of an object category, it is possible to distinguish three kinds of generalization:

(i) Inclusions between convex regions of domains.

For example, the regions associated with *terrier* are subregions of the regions associated with *dog*, and thus *dog* is a generalization of *terrier*.

(ii) Decreasing the number of meronomic relations specified for a category.

For example, *mammal* has four legs as part of its meronomy, but the more general *animal* does not.

(iii) Decreasing the number of domains associated with a category.

For example, a *plastic duck* may only share the shape domain with the biological concept of *duck*. In this way, the use of *duck* is generalized to "duck-shaped objects."[19]

Thus the sets of domains, their regions, and the meronomic relations involved in the characterization of a category generate a type hierarchy among object categories. Consequently this hierarchy does not have to be specified as part of the meanings of categories, but it is implicit in the structure of the categories. The importance of this point will be clearer when I discuss the limitations of the Semantic Web in chapter 14.

Let me bring out some immediate consequences of the generalization criteria. First of all, the *hierarchy* of generality of categories emanates as *logical* relations of generality of nouns.[20] For example, "An abyssinian is a cat" will automatically be true by the fact that the regions associated with the domains for the category of an abyssinian are subregions of those associated with the category of a cat.[21]

Second, *similarity* relations between categories explain why certain nouns are *synonyms* or near synonyms. As mentioned previously, Langacker (1987, p. 57) notes that "roe" and "caviar" both have the same extension—a mass of fish eggs. However, the prominence of some of the biological domains is higher for "roe" than for "caviar," while conversely the prominence of domains related to food is higher for "caviar" than for "roe." Similar analyses can be applied to the pairs "shore" versus "beach" and "ground" versus "land."

As part of prototype theory, Rosch (Rosch, 1975, 1978; Mervis & Rosch, 1981) introduces the *basic level* of a hierarchy of object categories as a particularly salient level of generalization. She presents a number of criteria for what distinguishes the basic level from superordinate or subordinate levels (see also Lakoff, 1987). One criterion says that superordinate categories contain much fewer common attributes (domains) than the basic level, and the subordinate levels contain hardly any additional common attributes (domains). For example, "cat" has many more characteristic attributes than "mammal," but not many more features than "abyssinian." Given the three kinds of generalization described here, the distinctions between different levels of a category can be made more precise. For example, subordinates to the basic level do not add new domains and seldom new meronomic relations. In general, subordinates narrow down the regions associated with the basic level. For example, an abyssinian has long ears and slender legs (proportion features of part-whole structure) and a "ticking" coat coloration (subregion of coat texture for cats). In support of this analysis, Hunn (1976) has argued that the basic level is the only level at which category membership can be determined by an overall *configurational* gestalt perception.[22]

6.5   Objects

6.5.1   Objects as a Special Kind of Categories

So far, I have focused on using conceptual spaces as a tool for modeling object categories (with properties as a special case that will be developed in the following chapter). However, it is also necessary to consider how single *objects* (or *individuals*, in the philosophical sense) can be represented in this framework. A straightforward way to represent an object is as a *point* in a conceptual space. Such a point can be seen as a vector of coordinates, one for each dimension in the space. In this way, each (physical) object can be allocated a specific spatial position, color, weight, shape, size, temperature, and so on. Two physical objects that seemingly share all properties (e.g., copies of the same book or plates in your cupboard) can still be distinguished on basis of their spatial location.

A consequence of this construction is that an object will always have an internally consistent set of properties—since, for example, "blue" and "yellow" are disjoint properties in the color space, it is not possible that any object will be both blue and yellow (all over). Consequently there is *no need for meaning postulates* (in the sense of Carnap) to exclude such contradictory properties.[23]

Recall that the vectors are cognitive construals; they do not necessarily refer to real objects. For example, one can see (or imagine) a horse without perceiving or representing its smell. In general, the construals will therefore not contain all the properties of an object. A more general representation that accounts for such situations is to use partial vectors, that is, points where the arguments for some dimensions are undetermined (cf. the discussion of fictional objects hereafter).

Representing objects as vectors (or partial vectors) in a conceptual space means that they are treated as a special kind of categories, that is, as categories where all regions (or known regions) of the domains are reduced to points. Thus objects can be seen as just very narrow categories.[24] This is the cognitive foundation for the fact that nouns, referring to categories, and names, referring to objects (individuals), have similar semantic (and syntactic) functions. As I argue next, though, there are further aspects of objects that distinguish them from categories.

6.5.2   Fictional Objects

Language does not only exist for talking about what there is. More important, it can be used to communicate about situations and objects that will be (for planning), things that are no longer (for sharing memories), and fictional objects (for story telling).[25]

If we assume that an object is completely determined by its set of properties (Leibniz's principle), then all points, specified for all domains in a conceptual space, can be taken to represent *possible objects*.[26] In this account, a possible object is a *cognitive* notion that need not have any form of reference in the external world.[27] Classical examples are centaurs, unicorns, trolls, and witches. By specifying a number of properties, one can create *fictional objects* represented as points in a conceptual space. Among other things, fictional objects are important for understanding the meaning of *quantifiers*, as I argue in chapter 12.

For most of the objects we represent in our conceptual spaces, we have only partial information. For example, for very few of the objects in my immediate environment do I know their weight or their smell, let alone their taste. My mind fills in with *default values* for these domains (based on my expectations). Consequently there is little cognitive difference between an existing object of which I have partial knowledge and a fictional object. This is, I submit, the reason why language hardly distinguishes between how real and fictional objects are described (see also Fauconnier, 1984).[28]

Readers with a philosophical background may think of my account of possible objects as similar to that proposed by Meinong (1960). A similarity is that objects can be identified with the aid of sets of properties. The main difference is that Meinong's theory allows round squares and other impossible objects. If an object is identified with a point in a conceptual space, however, it will at any point in time be assigned a consistent set of properties. Because *round* and *square* will be represented as disjoint regions of the shape domain, an object can never have both properties. If we use the notion of domains in conceptual spaces, we thus avoid the problem of impossible combinations of properties that has been an enigma for Meinongian theories.

6.5.3   Connected Paths

A characteristic of physical objects is that their *spatiotemporal paths are connected*. In other words, an object does not make any abrupt jumps in space. For example, if you drop a coin and it rolls under the bed, you will not find it up on the shelf in the next second. Modeling this characteristic requires that the space domain is always among the domains representing physical objects, and the trajectory in space of the object forms a connected path.[29] A consequence is that an object is represented mentally as having a unique location even if it is not observed at the moment. This characteristic of object representation is, fundamentally, what lies behind the psychological phenomenon of *object permanence* (Piaget, 1972). It also shows that it is not sufficient to identify objects with their sets of object category properties.

It may happen that two entities that were represented as distinct turn out to be identical. For example, the morning star was thought to be different from the evening star. It was an empirical discovery that their spatiotemporal paths are identical and that there is therefore only one object (the planet Venus).[30] And, the other way around, the woman you thought you saw riding on a bike yesterday turned out to be her twin sister.

Physical objects have other cognitive characteristics that are relevant for their identification:[31] they *occupy* space (so that two objects are never at the same place at the same time[32]), they are *bounded* in space, they *move independently* from spatially separated bodies, and they obey *force laws*. Furthermore, they exhibit some form of *cohesion* between their parts, as was discussed earlier. Clouds and shadows are therefore not good examples of objects, since their cohesion is weak and their spatiotemporal connectedness is uncertain.

6.6   Levels of Abstraction in Object Categories

The human mind deals with many kinds of entities, not just physical objects. Lyons (1977, pp. 442–445) distinguishes three fundamental "orders": (1) physical objects; (2) events, processes, states-of-affairs; and (3) propositions, schemas, and so on, that "are outside space and time" (p. 443).[33] On the basis of the spatial and temporal domains, this classification can be explained. A physical object always has a unique location in space and time, and it has a connected spatiotemporal trajectory.[34] An object is identified via its unique location in space at any moment of time. An object exists for a continuous period of time, often undetermined, but can change location. Thus for object categories it is natural to ask, for example, "Where is the moose," but not "*When is the moose?" (Langacker, 1991b, p. 14). An event (or a process or a state-of-affairs) also has a location in space and time (cf. Paradis, 2005, p. 551). In contrast to objects, events may occur at several places (an election, for example), but they are bounded in time. The identity criteria for events are dependent on the participants in the event, their interactions, and the time span. These criteria show interesting differences from those of objects (Lyons, 1977, p. 444). One can say, "The same moose was in the garden again today," but one cannot say, "*The same storm destroyed the garden again today." A category of events that is not covered by Lyons's classification comprises those that are not bounded in space, but only in time, for example, "summer." Summer spans a region of the domain *annual cycle*, and the identity criteria are more like those for physical objects: one can say, "Summer is here again."[35] Finally, Lyons's third-order entities have a location neither in space nor in time, and their identity criteria are therefore different. The upshot is that, by considering the different roles of the space and time domains, a coarse ontology of entities can be generated, including the three orders proposed by Lyons (1977).[36]

The three fundamental orders can be further divided. For example, Paradis (2005, pp. 549–553) proposes a division of physical objects into people, animals, plants, artifacts, natural objects and phenomena, locations, and substances. All these subdivisions are expressed by count nouns except for substances for which mass nouns are used. Ravid (2006) presents a noun scale from concrete to abstract divided into ten levels:

(i) Concrete nouns (e.g., ball, bike, jar, ruler)
(ii) Proper names (Oscar, Victoria, Alice)
(iii) Collective/location nouns (library, city, club, family)
(iv) Role nouns (driver, guide, mother, husband, employer)
(v) Generic nouns (kids, stuff, things, everybody)
(vi) Temporal nouns (minute, hour, month)
(vii) Event nouns (battle, lunch, session, therapy)
(viii) Imageable abstract nouns (blow, kick, answer, exam paper)
(ix) Abstract nouns (authority, dignity, harmony, opinion)
(x) Derived abstract nouns (communication, discussion, intervention, separation)

Ravid (2006, p. 798) says that levels (i) and (ii) correspond to Lyons's first order, levels (iii)–(v) are transitions to the second order, levels (vi)–(viii) are second order, and levels (ix) and (x) are third-order entities.

On the basis of the many kinds of nouns, Jackendoff (1994, pp. 68–69) argues that not every noun names an object. Langacker (2008, chap. 4) defends the position taken here. He argues that abstract objects can be created by reprofiling of image schemas. For example, he mentions that the image schema for the verb "choose" can be reprofiled in three ways, resulting in three meanings of the noun "choice": (i) the element chosen ("her top choice"), (ii) the range of alternatives to choose from ("they offer a wide choice"), and (iii) the act of choosing ("she made her choice quickly").

6.7   Relational Categories

"I've often seen a cat without a grin," thought Alice; "but a grin without a cat! It's the most curious thing I ever saw in all my life!"
—Lewis Carroll

Gentner and Kurtz (2005) discuss *relational categories*, for example, *ally*, *passenger*, and *robber*, that are determined by a common relational structure, rather than by properties in a domain matrix. They note that research on object categories has mostly ignored relational categories. Although Gentner and Kurtz note that other word classes also include relational concepts, their focus is on nominal relational categories. Even so, it is a rather scattered class of nouns they describe as relational: bridge, carnivore, passenger, accident, investigation, mistake, payback, and so on. They also argue that relational categories, in contrast to object categories, are, in general, linguistically constructed (Gentner & Kurtz, 2005, p. 156).

I cannot provide a complete analysis of all relational categories, but I will focus on nominalizations that are generated from verbs. Let me take the example "robber." The verb root for this relational category is "rob," which is subsumed under "steal." The events that these verbs describe involve (at least) three roles: someone who steals (agent), someone who is stolen from (patient), and the goods stolen (patient/theme).[37] Each of the roles is expressed by relational nouns: for the first one, "thief" or "robber" (for the agent of a subcategory of events involving personal encounter between the one who steals and the one stolen from); for the second, "victim" (a noun that can be used for a broader category of events); and for the third, "goods" (a noun that can also be used for a broader category of entities). In brief, the category *robber* is defined by its role in an event of *robbing* and not just by properties from category space.[38] The category *robber* contains objects (Lyons's order [1]) that can be the agents of a category of events. On a more abstract level, the entire event construal, involving all three roles, is described by *robbery* (Lyons's order [2]). I analyze the structure of events in chapter 9.

This ends my presentation of how object categories are represented in conceptual spaces. It is now time to connect to object categories to words.

6.8   The Semantic Function of Nouns

6.8.1   Nouns and Names
The linguistic category of *nouns* seems to occur in all languages. From a cognitive point of view, an obvious question is why this word class is so

central. If we look at the role of nouns in communication, there is a rather straightforward answer: the basic function of a noun is to identify a *referent* (as an instance of a concept). On the first level (instructions), using a noun is helpful in identifying an object that can be used in cooperation. Wittgensteinian language game examples such as "d–slab–there" are typical. In this case, the referent (the slab) is normally present in the communicative environment. On the second level (coordination of common ground), the speaker may use a noun to introduce a new referent to the common ground or to inform about the properties of an already established referent. In these cases, the referent is typically *not* present in the communicative environment. Being able to share nonpresent referents is crucial for *planning for future cooperation*. In section 3.4, I argued that this capacity is a central factor in the evolution of symbolic communication. On the third level (coordinating meanings), nouns are typically used in generics to express properties of categories ("A spider has eight legs") or to express relations between object categories ("Spiders are not insects").

Informal definitions of nouns say that they refer to persons, places, things, events, substances, and so on. However, linguists are in general not satisfied with this kind of description, since it is difficult to specify a complete list of categories. Furthermore, the definition uses abstract nouns to define the very notion of a noun, which makes it slightly circular.

Given the previous analysis, one can get out of this circle by formulating the following basic semantic thesis:

Thesis about nouns: Nouns refer to concepts.
Thesis about names: Names refer to objects.

Here a name is seen as a special kind of noun with a more direct reference to a particular object (individual). Although the theses may seem trivial, they gain a rich content from the analyses of concepts and objects that I presented earlier in this chapter.

The ontologically (and developmentally) primary references of nouns are object categories.[39] As indicated in the previous section, there are several other ontological categories that are denoted by nouns.

Now, if the primary communicative function of a noun is to express a referent and names identify a unique referent, why does not everything have a name? It would seem that if every object had a name, this would eliminate a large number of ambiguities. The reason why nouns are needed is one of *cognitive economy* (see, e.g., Kemp & Regier, 2012). Our memory is severely constrained, and it would be impossible to learn and remember a name for all objects one wants to communicate about. Nouns referring to basic-level categories group the objects in the world in categories that are suitably large for communication, so that ambiguities become sufficiently rare. The cognitive solution is a balance between the precision of the noun and the number of words that have to be remembered. The precision of a noun is partly determined by the mechanisms for vagueness (discussed in sec. 2.8) and partly by context. As I argue in chapter 13, a basic method of achieving further precision to solve a problem of uniquely identifying a referent is to use combinations of words.

The nouns first learned by children mainly refer to concrete objects (Gentner & Boroditsky, 2001). Markman (1992, 1994) and Bloom (2002) point out biases that are involved in the learning of nouns. One is *a bias against lexical overlap*. For example, if a child is shown two objects—a rabbit and a strange, novel machine—and the adult says, "Point to the fendle!" the child will most likely point to the machine (if it already knows the meaning of "rabbit"). Bloom (2002, pp. 42–43) argues that this bias toward mutual exclusivity of the meaning of words derives from children expecting speakers to be informative (a Gricean principle), so they expect familiar things to be named by familiar words.

Another bias is the *whole object bias*. Independently of linguistic cues, children are prone to attend to whole objects, and as a consequence, they assume that they are salient to others as well (Markman, 1992). Even babies can track objects and distinguish whether there is one or several objects (Spelke, 1994). So when an adult uses a new noun and her attention has a certain direction, the child will assume that the reference of the word is a salient object in that direction.

6.8.2 Mass Nouns and Count Nouns

An important semantic distinction is that between *mass nouns* and *count nouns*. This distinction has been accounted for in meronomic terms (Ojeda, 1993; Moltmann, 1998, p. 79). Say that x is an *atom* in a part-whole structure if x has no (proper) parts. Then one can formulate the criterion that if N is a *mass noun*, then the references of N (the entities falling under the concept) contain no atoms. However, this criterion is not unproblematic, since it implies that mass nouns denote indefinitely divisible quantities. This may cause problems for some mass nouns, for example, "furniture." One solution to this problem is to argue that the mental construal does not contain atoms, but that is putting the cart in front of the horse.

Moltmann (1998) argues that an extensional account of meronomic structure is not sufficient to explain the mass-count noun distinction, but the notion of an *integrated whole* must be added. She writes that "when count nouns are converted into mass nouns, generally an implication of integrity gets lost in the process. Thus *apple* as a count noun implies a certain shape, whereas *apple* as a mass noun rather suggests the loss of shape (for example, pieces of apple)." The problem then is, of course, to explicate the notion of an integrated whole. Here Moltmann has no definite answer, but she claims that the prototypical way for an entity to be an integrated whole is by having a particular shape. However, this proposal also has problems: a cloud or a crowd has a shape but is not an integrated whole. I have no definite proposal for how to describe integration, but maybe a better account is to include the *force relations* between the different parts (integrated wholes *hang together*).[40]

Another way of accounting for the mass-count distinction is to say that for mass nouns, the *material domain* becomes the most prominent one; the shape domain is of minor importance. For example, "apple" as a mass noun just concerns the material. Furthermore, the constraint of spatiotemporal connectedness of the paths of objects does not apply to mass nouns. The water in the bucket can be divided into several amounts of water without any identity being lost. This concludes my presentation of the relations between concepts and nouns. Further aspects of nouns will be brought up in discussions of their relations to adjectives in the following chapter and to verbs in chapter 10.

7 Properties and the Semantics of Adjectives

7.1 Adjectives and Domains

Simply put, nouns (and names) refer to objects, and verbs say something about what happens to objects. So why do we have adjectives in language?[1] It is useful to return to the three levels of communication introduced in section 5.1. On the first level (instruction), an adjective is used as a *specification* of a noun (or noun phrase) that helps in identifying a referent. For example, if you want somebody to fetch you a particular book and there are several books present in the context, you specify it further by saying "the green book" or "the big book." In addition to this kind of "element identification," Frännhag (2010) also considers "kind identification." For example, you may request "a thick book" when you want something to put on a chair for a child to sit on, but it does not matter which book is delivered.[2] Here the goal is not a particular object but something that has a desired property.

On the second level (coordination of common ground), a typical function of an adjective is *informative*. You can say, "The stove is hot," as a warning to somebody. Linguistically, the adjective is then a complement to a copula ("is") or an intransitive verb ("the meal tastes wonderful"). The specification function and the informative function may very well be cognitively separated. Therefore it is not obvious that these functions should be expressed by one word class. Dixon (2004, p. 30) notes that some languages have two different word classes, one fulfilling the first function and another fulfilling the second. Actually, some words classified as adjectives in English only have one of the functions: "Afraid" and "alive" can only be used informatively (predicatively), and "absolute" and "main" can only be used as specifications (attributively) (Paradis, 2005). The specification function can also be fulfilled by a noun. For example "the silk scarf" can be used to distinguish among several scarves. On the third level (coordination of meaning), an adjective is typically used in *generics* ("cloudberries are orange") to specify a property characteristic of an object category (or some other concept).

I base my account of the semantics of adjectives on the analysis of properties in section 2.2. The key idea is that adjectives express properties. This generates the following thesis:

Single-domain thesis for adjectives: The meaning of an adjective can be represented as a convex region in a single domain.

For many adjectives, the single-domain thesis seems to be valid, in particular for adjectives that relate to domains that are acquired early in language learning. In particular, I conjectured (Gärdenfors, 2000) that all color terms in natural languages express convex regions with respect to the color dimensions of hue, saturation, and brightness. This means, for example, that no language should have a single word for the colors denoted by "green" and "orange" in English (and which includes no other colors), since such a word would represent two disjoint areas in the color space. I present strong support for this conjecture in the next section.

Adjectives are also used for *comparing* things: many languages have comparatives, such as "taller" and "smarter," which can be used both to specify ("the taller woman") and to inform ("Victoria is smarter than Oscar"). Many languages also have superlatives, for example, "tallest" and "smartest," which can be used in like fashion to comparatives.

Nouns seldom allow comparisons (Schwarzschild, 2008, p. 319): expressions like "*dogger" and "*more dog" are hardly used. Why do nouns not allow comparatives, when adjectives do? A comparison requires a well-defined dimension along which the comparison is made. Since many adjectives are based on one-dimensional domains, a comparative (and a superlative) then simply involves a comparison of the values along this dimension. For example, "Oscar is taller than Victoria" means that Oscar's coordinate on the length dimension is greater than Victoria's.

If the adjective is based on a multidimensional domain, the situation becomes more complicated. In general, the comparative means "closer to the prototype for the adjective": "Greener" means closer to the green prototype in color space; "healthier" means a lower value on some of the dimensions involved in illness (see sec. 7.3).[3]

7.2   Evidence for the Convexity Thesis: Color Properties

The single-domain thesis for adjectives was proposed in Gärdenfors (2000). I used color words as the prime example, since the geometry of the color space is relatively well explored. At the time, I had limited empirical support for my claim. Taft and Sivik (1997) (see also Sivik & Taft, 1994) had studied a few languages (Swedish, Polish, Spanish, and American English) and found that the color words in these languages do indeed correspond to convex regions of the color space.

Jäger (2010) tested the prediction extensively. He investigated all the data from 110 languages that are included in the World Color Survey project (for details, see Cook, Kay, & Regier, 2005). During the collection of the data, the 330 Munsell color chips were used. They cover 322 colors of maximal saturation, located on the surface of the double cone of colors, and eight shades of gray from white to black on the central axis of the double cone.[4] Note that the colors located in the interior of the double cone are not included in the set of Munsell color chips. Thus the chips do not generate complete information about how color words are used in different languages.

For each language and for each pair of color terms, Jäger calculated an optimal linear separator. This means identifying a plane that cuts through the color space and separates as many of the chips from the color categorization as possible. Taking the intersection of all the linear separators for a given color word generates a convex region of the color space. He then calculated the proportion of Munsell chips that are not reclassified by the linear separator, which is a way of measuring how well the color categorization fits the convexity assumption. Due to limitations in the data, only 102 languages were considered, but Jäger found that the mean value over these languages is 93.8 percent. This is an extremely high value, since there are always statistical aberrations in the data, in particular for languages within the World Color Survey project with data from only a few informants. The upshot is that the convexity thesis for properties receives very strong support from color adjectives in a large variety of languages.[5] Obviously this does not give any immediate support for the convexity thesis with regard to other domains. Each domain will require its own empirical evidence (unless arguments based on metaphorical mappings can be recruited).

7.3   Evaluating the Single-Domain Thesis

Since I have not presented any sharp criteria for what can count as a domain, at least not for abstract properties, the reader is justified in asking whether the single-domain thesis for adjectives can be *falsified*. One example of an adjective, the meaning of which has been proposed to involve several domains, is "healthy." Being healthy means being at one end of a lot of dimensions: not having pain, not having a high temperature, not having an infection, not having high blood pressure, not having a high cholesterol level, and so on. So the meaning of "healthy" does not seem to satisfy the single-domain thesis.

There are two routes out of this dilemma. The best one, in my opinion, is to say that the product space of all the dimensions relevant for "healthy" forms an illness–health domain. In support of this, one can argue that this domain is used by doctors when categorizing different forms of illnesses, so that it is a source of regions for a large variety of concepts. "Healthy" would then just correspond to a corner region of this domain. The drawback of this route is that it does not tell us which domains can be used to form product domains, so it does not completely solve the nonfalsifiability problem. The other route, which I find less attractive, is to say that one can identify a "health dimension" that mathematically could be described as the "diagonal" in the product space of all the dimensions involved in health and illness. Again the problem would arise of when such diagonal dimensions can be construed.

Even if it turned out that the research program of identifying relevant domains would not develop in the direction predicted by the single-domain thesis for some classes of adjectives, the thesis still has value, since it works well for a large majority of adjectives, in particular the adjectives that are among the first to be learned by children. A weaker version of the single-domain thesis could still be a heuristic for how the meaning of adjectives is determined, and this heuristic makes it easier to learn what they mean (cf. Bloom, 2000, chap. 8, on syntactic markers as tools for learning language).

7.4 Classifications of Adjectives

In the literature, one finds a large number of classifications of adjectives, for example, Dixon (1977, 2004), Bierwisch (1987), Rachidi (1989), Hundsnurscher and Splett (1982), Lee (1994). Although the classifications vary considerably, they largely support the single-domain thesis. For most of the classes, it is straightforward to identify the underlying domain. To give a single example, Dixon's (2004, pp. 3–4) list of semantic types typically associated with adjectives contains four central types, *dimension* (referring to adjectives based on the visuospatial domain), *age*, *value* (good, bad), and *color*; and three peripheral types, *physical property* (referring to properties of objects such as *heavy* and *hard*), human propensity (jealous, happy, clever, cruel, etc.), and *speed*.[6] This list fits well with the analysis in chapter 3 demonstrating that the object category (of which color is a subdomain), spatial, time, and value domains all are among the earliest in children's development.

But the classification also includes a couple of classes that are more difficult to evaluate with respect to the single-domain thesis. For example, the class *human propensity* contains "friendly" and "jealous." These adjectives involve relations between humans, and it is not easy to see what the underlying domains are (although "jealous" presumably involves the emotion domain as one of its components in some product space also involving goals).

Blackwell (2000) studied how children acquire syntactic structures for adjectives in relation to their specification and informative functions. She found that the adjectives from Dixon's dimension, age, value, color, and speed types tend to be used first as modifiers (specification function), while those from the physical property and human propensity types tend to be used first as copula complements (informative function). This finding corroborates the thesis that the two communicative functions of adjectives are cognitively separated.

However, Dixon's list is not exhaustive; it omits many abstract types of adjectives. It is difficult to provide a list of levels of abstraction in adjectives that could parallel that of nouns presented in the previous chapter. One division is proposed by Paradis (2005, pp. 556–560). She distinguishes between *content-biased* adjectives, such as those presented in Dixon's (2004) list, and *schematicity-biased* adjectives, such as "absolute," "frequent," "main," and "possible." Paradis (2005, p. 555) says that schematicity-biased adjectives refer to *shells*, which in her terminology are third-order entities and presumably of a more abstract nature than domains. Thus, lacking an analysis of shells in terms of domains, the single-domain thesis for adjectives should perhaps be restricted to content-biased adjectives.

Furthermore, the class of adjectives can be extended indefinitely by derivations based on verbs and nouns, for example, "growing," "screaming," "woolen," "female." For such derivations, it may be difficult to identify a unique corresponding semantic domain. The upshot is that, for more abstract classes of adjectives, the single-domain thesis must be seen as a research program that requires the theory of domains to be developed in further areas.

7.5   Scalable and Nonscalable Domains

When evaluating the single-domain thesis, it must be noted that domains come with varying structures (see, e.g., Bierwisch, 1967; Paradis, 2001).

First, some domains, such as gender, are merely *classificatory*: individuals belong to a small number of classes. Second, many domains, such as length and weight, are *scalable*: they are determined by a single dimension. Individuals have values along this dimension. The scalable domains can be subclassified by whether the dimension is open-ended or whether it has one or two endpoints (boundaries). *Time* is a dimension that is open-ended in both directions. *Height* and *weight* both have a lower bound (zero) but are open-ended upward. The dimension of fullness has two endpoints: empty and full. Finally, some domains, such as the color, health, and emotion domains, have *multiple* dimensions.

Following Paradis (2001, 2008), I next want to show how the geometric structure of a domain influences the linguistic possibilities of the adjectives that refer to the domain. In particular, the scalable domains allow for several constructions that are not available for nonscalable domains and vice versa. A first example is *comparatives*. Classificatory domains do not allow comparatives: you are either dead or not; there is no "*deader" or "*deadest." Scalable domains generate comparatives in a highly straightforward way: an object x is *longer* than another object y if the value of x on the length dimension is larger than that of y. An object x is *longest* if no other object y has a larger value on the length dimension than x.

As noted previously, comparatives can also be used for adjectives that refer to domains with multiple dimensions: "redder," "reddest." Here the semantic mechanism is different, however: "x is redder than y" means that x is *closer* in the color domain to the prototype of red than y is. Similarly, "x is reddest" means that x is the individual (in a contextually given class) that is *closest* to the prototype. Notice that for this domain, closeness is determined in a space of three dimensions.

A second example concerns the types of *adverbial expressions* that combine with the different kinds of domains. For example, Paradis (2001, p. 50) considers scalar adverbs such as "very," "terribly," "fairly," and "slightly." They combine easily with adjectives from scalar domains, but only in marginal language uses with adjectives from classificatory domains: "?very dead," "?fairly male." For scalar domains with endpoints, there are *totality adverbs* such as "completely," "absolutely," and "almost" that mark nearness to an endpoint: "completely full," "almost ripe" (pp. 50–51). The totality adverbs do not combine with open-ended scalar domains: "*completely long," "*absolutely warm." The totality adverbs can be used with multidimensional domains, but again, they mark nearness to a prototype: "absolutely blue," "completely healthy."

Paradis (2008, p. 331) points out that "boundedness" is strongly tied to gradability and is important in many construals. She writes that "crosscat-egorial correspondences have been recognized between count and non-count structures in nouns (*car, mistake* vs. *milk, information*), and continuous and noncontinuous structures in verbs (*know, hate, play* vs. *arrive, die, cough*) . . . and adjectives (*good, long* vs. *dead, identical*)" (see also Langacker, 2008).

The upshot is that the topological structure of the domain that is associ-ated with an adjective (or a class of adjectives) will determine which adverbial modifiers it can be combined with. Conversely, the adverbs used with an adjective reveal the assumed structure of the underlying domain (see, e.g., Tribushinina & Gillis, 2012). I return to an analysis of different kinds of adverbs in section 12.3.

An interesting case of modifiers for adjectives comes from the color domain. This domain is built up from three dimensions: hue, intensity, and brightness. It turns out that adjective modifiers affect each of these dimensions: "bluish green" marks a shift in hue, "pale green" a shift in intensity, and "light green" a shift in brightness.[7]

7.6   Developmental Aspects of Properties and Adjectives

For young children, adjectives are more difficult to learn than nouns. In the terminology of chapter 3, the reason seems to be that to learn an adjec-tive, the child must have separated out the domain that the adjective refers to. Smith (1989, p. 159) presents a closely related analysis:[8]

There is a *dimensionalization* of the knowledge system. . . . Children's early word acquisitions suggest such a trend. Among the first words acquired by children are the names for basic categories—categories such as *dog* and *chair*, which seem well organized by overall similarities. Words that refer to superordinate categories (e.g., *animal*) are not well organized by overall similarity, and the words that refer to dimensional relations themselves (e.g., *red* or *tall*) appear to be understood rela-tively late.

A related phenomenon from child language is that adjectives that denote contrasts within one adult domain are often used for other domains as well. Thus three- and four-year-olds confuse "high" with "tall," "big" with "bright," and so on (E. Clark, 1972; Carey, 1985). This indicates that the domains are not yet sufficiently separated in the minds of the children.

The hypothesis that adjectives are more abstract tools for communica-tion than are names and nouns is supported by data from child language. One example comes from Mintz and Gleitman (2002), who showed that if the adjective comes together with a noun that already has a name, then even two-year-olds can learn new adjectives quickly (see also Waxman & Markow, 1998). Mintz and Gleitman (2002, p. 285) conclude that "24- and 36-month-olds do not seem to map novel adjectives to object properties without the support of a full noun."

In section 3.3, I presented the establishment thesis, which says that if one word from a domain is learned during a certain establishment period, then other words from the same domain should be learned during roughly the same period. By definition, properties are regions of single domains. Therefore the thesis can be tested efficiently by studying adjectives that refer to the same domain. A simple way of finding adjectives referring to the same domain is to study *antonyms*. Mintz and Gleitman (2002, p. 273) note that "antonymous adjectives are far more errorlessly and early-acquired than the multivalued nonpolar adjectives." Even if the identification of domains may sometimes not be clear, an antonym pair is certain to refer to the same domain (at least to the extent that the adjectives do not have multiple meanings). Table 7.1 contains a selection of adjective antonym pairs and the establishment periods for each of the two words.[9]

As can be seen from table 7.1, a strong correlation exists between the establishment of an adjective and its antonym. The only obvious exception is *tight–loose* (which may be due to idiomatic uses of "tight," such as "sleep tight"). There is also a clear tendency for the "positive" (unmarked) adjective in the pair to appear first.[10] Furthermore, note that domains can be divided roughly into two sets: one basic set that is established around twelve to eighteen months, and another, containing more abstract domains, around thirty-six to forty-two months.[11]

Deese (1964) asked subjects to associate dimensional adjectives with words. The most frequent word elicited was very often an antonym, although sometimes more than one antonym occurred. He then calculated the factor loadings between eleven common pairs of antonyms. The result was that, with the exception of the pairs "large–small" and "big–little," the dimensions expressed by the pairs were almost orthogonal. Since "large–small" and "big–little" both refer to the visuospatial domain, it is no surprise that they are correlated. That the other dimensions are almost orthogonal supports that they are separable and thus represent distinct domains.[12] The factor analysis used by Deese (1964) can potentially also be used to investigate the integrality and separability of more abstract dimensions.

ChildFreq is, however, a blunt instrument, which forces the selection of adjectives in table 7.1 to be slightly biased. For example, many pairs like

Table 7.1
Establishment periods from ChildFreq for adjective antonyms: age range from 6 to 71 months, 6-month intervals, number given for the end of the establishment period. * = several antonyms, + = positive adjective, – = negative adjective.

Adjective	Est. period	Antonym	Est. period
hot	12	cold	12
old+*	12	young–	12
clean	12	dirty	12
dry	12	wet	12
old+*	12	new–	12–18
big+	12	little–	12–24
tall+	12	short–	18
heavy+	12	light–*	18
sad–	12	happy+	18
loud+	12	quiet–	18
empty–	12–18	full+	18–24
high+	18	low–	18
light*	18	dark	24
sweet	18	sour	24
tight+	18	loose–	48
right	24	wrong	24
curly	30	straight	30
first	30–42	last	42
expensive+	36	cheap–	36–42
fat+	36	skinny–	42
wide	36	narrow	36
thick+	42	thin–	54

"rich–poor" and "easy–difficult" cannot be evaluated because "poor" and "easy" have other uses, and for many adjectives, the frequency of use is too low to be evaluated. Nevertheless the data provide further support for the establishment thesis from section 3.3. The data deserve to be examined with sharper methods whereby the corpus is examined more carefully.

The central semantic thesis of this chapter is that the meaning of an adjective is represented by a convex region in a single domain. The thesis forms the basis for many testable predictions concerning the semantics of adjectives. Here I have only presented a few, but I invite linguists and cognitive scientists to further develop and test the predictions. A limitation of the thesis, however, is that for some classes of adjectives, in particular abstract ones, it may be difficult to identify the underlying domain and to pin down its geometric or topological structure.

8 Actions

8.1 Actions Based on the Force Domain

The previous two chapters have analyzed the semantics of nouns and adjectives on the basis of concepts (mainly object categories) and properties respectively. Object categories and properties are static concepts. It is obvious, however, that a considerable number of our cognitive representations concern *dynamic* properties (see, e.g., van Gelder, 1995; Port & van Gelder, 1995).[1] Consider for a moment what verbs represent in natural languages: typically, verbs express dynamic properties of objects, in general as parts of events. Since events normally involve actions, I will prepare the ground for the analysis in the following chapters by providing a model of actions in terms of the force dimension.

I argue that the *action domain* can be analyzed in the same way as, for example, the color domain or the shape domain. My starting point is that the categorization of actions depends, to a large extent, on the mental representation of *forces*. My core theses are that an action can be described as a *pattern of forces*, and an action category can be described as a convex region of such patterns.

In the tradition of Lakoff and Langacker within cognitive linguistics, the *spatial* structure of the image schema has been in focus (the very name "image" schema indicates this). Lakoff (1987, p. 283) goes as far as to put forward what he calls the "spatialization of form hypothesis," which says that the meanings of linguistic expressions should be analyzed in terms of spatial image schemas plus metaphorical mappings. For example, many uses of prepositions, which primarily have a spatial meaning, are seen as metaphorical when applied to other domains (see, e.g., Brugman, 1981; Herskovits, 1986).[2]

The psychological and philosophical literature on concepts has also tended strongly to focus on concepts that are grounded in perceptual mechanisms. For example, the theory of conceptual spaces developed in Gärdenfors (2000) deals mainly with perceptual concepts based on dimensions such as color, size, shape, and sound. That said, strong evidence exists that many of our everyday concepts derive from our understanding of actions and events. Furthermore, most artifacts—such as chairs, clocks, and telephones—are categorized on the basis of their *functional* properties (Nelson, 1996; Mandler, 2004), that is, what you can *do* with them.

Actions and events also seem to be central to the development of concepts in infants and children. Nelson (1996, pp. 110–111) summarizes:

> The claim was that infants would form concepts of objects that were functional in their own lives, and would therefore learn to name them easily and early. The object's function from the child's point of view (e.g., rolling a ball) would form the core of the concept, and its perceptual appearance would be added as a cue to the identification of instances. Later we proposed . . . that object concepts were first embedded in event representations, accounting for some of the earliest uses of object words to refer to whole events. . . . For example, goals and locations are critical aspects of events.

Langacker's analysis of image schemas is presented mainly in spatial terms. For example, his description of "climb" (see sec. 1.3.2) mentions only the vertical dimension, together with the time dimension. No forces are involved in his analysis. Thus the schema does not differentiate between "pull up," "push up," and "climb." What is missing is that the meaning of "climb" involves the trajector exerting a vertically directed force.

A force has a *magnitude* (it can be weak or strong), a *direction*, and a point of *origin*. Thus a force can be represented as a *vector*. In this and the following chapter, I want to show that the mathematical properties of vectors contribute to the understanding of the structure of actions and events. I will here give three examples of the benefits of vector representations. First, vectors can be more or less *close* to one another. As I will argue, this corresponds to similarity of actions (via the closeness of the corresponding force vectors or force patterns). In chapter 10, I show that closeness of vectors provides a natural explanation for similarities of verb meanings. Second, vectors can be *added* and *multiplied*. Wolff (2007, 2008) has shown that subjects' judgments of causality are dependent on such additions of forces and counterforces.[3] Third, sets of vectors can form *convex* sets.

The space of force vectors constitutes the *force domain*. By adding the force domain to those involved in object categories, we obtain the basic tools for analyzing dynamic concepts. The forces involved need not only be physical forces but can also be *social* or *emotional* forces (I return to this extension hereafter). As before, it is the cognitive representation of forces that counts, not physical theories.

Johnson (1987, pp. 45–48) presents a number of "preconceptual gestalts" for force. These gestalts function as the correspondence to image schemas, but with forces as the basic organizing feature rather than spatial relations. The force gestalts he presents are "compulsion," "blockage," "counterforce," "diversion," "removal of restraint," "enablement," and "attraction."

Another proponent of the dynamic perspective is Talmy (1988), who emphasizes the role of forces and dynamic pattern in image schemas in what he calls "force dynamics." He recognizes the concept of force in expressions such as the following:

(8.1)   The ball kept (on) rolling along the green.

(8.2)   John can't go out of the house.

Talmy also notices the possibility in language of choosing between what he calls force-dynamically neutral expressions and ones that do exhibit force-dynamic patterns, as in the following:

(8.3)   He didn't close the door.

(8.4)   He refrained from closing the door.

Talmy's dynamic ontology consists of two directed forces of unequal strength, the focal called "Agonist" and the opposing element called "Antagonist," each force having an intrinsic tendency toward either action or rest, and a resultant of the force interaction, which is either action or rest. The forces are furthermore taken as governing the linguistic causative, extending to notions like letting, hindering, helping, and so on. He develops a schematic formalism that, for example, allows him to represent the difference in force patterns in expressions like "The ball kept rolling because of the wind blowing on it" and "The ball kept rolling despite the stiff grass." A limitation of his model is that he considers only two opposing forces.[4]

8.2   Representing Actions by Patterns of Forces

8.2.1   Patterns of Forces
One idea for a model of actions comes from Marr and Vaina (1982) and Vaina (1983), who extend Marr and Nishihara's (1978) cylinder models of objects (see sec. 6.3.1) to an analysis of actions. Marr and Vaina's model

describes an action via differential equations for movements of the body parts of, for example, a walking human (fig. 8.1).[5]

It is clear that, using Newtonian mechanics, one can derive these equations from the *forces* applied to the legs, arms, and other moving parts of the body. For example, the pattern of forces involved in the movements of a person running is different from the pattern of forces of a person walking; likewise, the pattern of forces for saluting is different from the pattern of forces for throwing (Vaina & Bennour, 1985). Just as for shapes, so the space within which force patterns are located can be treated as an integral domain, with its own metric.

The human cognitive apparatus is not exactly built for Newtonian mechanics. Nevertheless I hypothesize that the brain extracts the forces that lie behind different kinds of movements and other actions. I present support for this hypothesis in the next subsection. This leads me to the following thesis:

Representation of actions: An action is represented by the pattern of forces that generates it.

An important consequence of this thesis is that the individuals or objects involved in the action are not part of the representation of the action. The following section presents empirical evidence to support this idea. I speak of a pattern of forces because, for bodily motions, several body parts are involved, and thus several force vectors are interacting (by analogy with Marr and Vaina's differential equations). The action domain is thus a *configurational* domain (see sec. 2.5.2) based on the force domain.

One can represent these patterns of forces in principally the same way as the patterns of shapes described in section 6.3. In analogy with shapes, force patterns also have *meronomic* structure. For example, a dog with short legs moves in a different way than a dog with long legs. Westera (2008) has developed a detailed model of force patterns for bodily motions as a hierarchy, where the lowest level is a *motor space* for each of the body joints. The next level is an *effector space*, constituted by the forces applied to a set of joints. To represent the simultaneous motor pattern of several effector systems—as when, in a tennis serve, one coordinates the throwing of a ball by one arm with a swing of the other arm—Westera introduces a *segment space*. On the highest level of the hierarchy, the segments are combined into a *motor space* representing the complete pattern of forces that generates an action. Westera's (2008) model is developed with robotic implementations in mind.

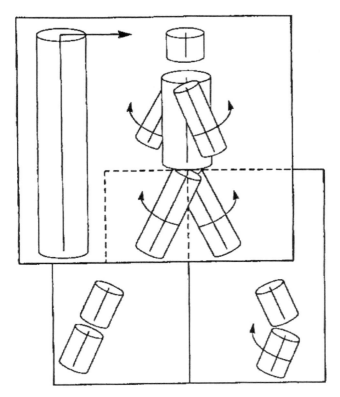

Figure 8.1
Walking, represented by cylinder figures and differential equations. Reprinted from
D. Marr and L. Vaina. (1982). Representation and recognition of the movements of
shapes. *Proceedings of the Royal Society in London B 214*, 501–524, by permission of
the Royal Society.

A theory that is related to the one presented here is that of Cohen
(1998). Instead of calculating the forces (which can be derived from second
derivatives of movement) involved in an event, he studies the *velocities*
(represented by first derivatives of movement) of the objects involved. He
shows that by systematically studying the qualitative patterns of the relative velocities of two objects, we are able to represent the basic meaning
of many simple verbs.

8.2.2 Empirical Evidence
The best empirical support for the thesis about force patterns comes from
psychophysics. Johansson (1973) developed a patch-light technique for

analyzing biological motion without any direct shape information. He attached lightbulbs to the joints of actors who were dressed in black and moved in a black room. The actors were filmed performing actions such as walking, running, and dancing. Watching the films—in which only the light dots could be seen—subjects recognized the actions within tenths of a second. Furthermore, the movement of the dots was immediately recognized as coming from the actions of a human being. Further experiments by Runesson and Frykholm (1981, 1983) showed that subjects extract subtle details of the actions performed, such as the gender of people walking or the weight of objects lifted (where the objects themselves cannot be seen).

Another example of data that can be used to study force patterns comes from Wang et al. (2004). They collected data from the walking patterns of humans under different conditions. Figure 8.2 shows the moments (Nm/kg) of the three leg joints, from which the force patterns can easily be derived (by calculating the derivatives with respect to time). Using the methods of Giese, Thornton, and Edelman (2008), we can use these patterns to calculate the similarity of the different gaits.[6]

One lesson to learn from the experiments of Johansson and his followers is that the kinematics of a movement contain sufficient information to identify the underlying dynamic force patterns. Runesson (1994, pp. 386–387) claims that people can directly perceive the forces that control different kinds of motion (see also Wolff, 2008). He formulates the following thesis:

Kinematic specification of dynamics: The kinematics of a movement contain sufficient information to identify the underlying dynamic force patterns.

Runesson (1994, pp. 386–387) further argues that one need not make any distinction between visible and hidden properties:

The fact is that we can *see* the weight of an object handled by a person. The fundamental reason we are able to do so is exactly the same as for seeing the size and shape of the person's nose or the color of his shirt in normal illumination, namely that *information* about all these properties is available in the optic array.

From this perspective, the information that the senses—primarily vision—receive about the movements of an object or individual is sufficient for the brain to extract, with great precision, the underlying forces. Furthermore, the process is automatic: one cannot help but perceive the forces. Of course, the perception of forces is not perfect; people are prone to illu-

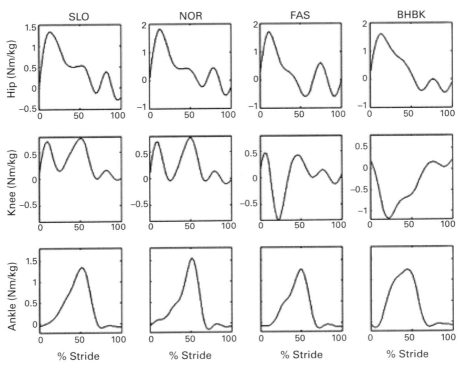

Figure 8.2

Joint moments for four different kinds of walking: slow (SLO), normal (NOR), fast (FAS), and bent-hip bent-knee walking (BHBK) for the hip, knee, and ankle joints. Reprinted from W. Wang et al. (2004). Comparison of inverse-dynamics musculo-skeletal models of AL 288-1 Australopithecus afarensis and KNM-W IS000 Homo ergaster to modern humans, with implications for the evolution of bipedalism. *Journal of Human Evolution*, *47*, 453–478, © 2004, with permission from Elsevier.

sions, just as in all types of perception. The thesis of kinematic specification of dynamics was formulated with respect to biological motion. I speculate that it extends to other forms of motion. That said, the empirical basis for such an extension is, so far, not strong.

It is obvious that the principle of kinematic specification of dynamics accords well with the representation of actions that I have proposed. One difference, though, is that Runesson takes a Gibsonian perspective on the available perceptual information, meaning that he would find it method-ologically unnecessary to consider mental constructions such as concep-tual spaces. From a Gibsonian perspective, the world itself contains sufficient information about objects and events so that the brain can just "pick up" that information, so as to categorize any entity (Gibson, 1979). I do not accept such a perspective. In my view, the perceptual information received is often incomplete, and consequently the brain must "construct" an interpretation.

Further empirical support for the proposed representation of actions comes from phonetics.[7] Browman and Goldstein (1990) describe the act of uttering a word as a "score of gestures" where the gestures are performed not by the hands but by the five vocal organs of velum, tongue tip, tongue body, lips, and glottis (see fig. 8.3 for some examples). They then describe the utterance of a word as a temporal sequence—a score—of activation of these organs. Each activation is meant to be one of a limited number of types. Such a score can be redescribed as a temporal pattern of force vectors. Browman and Goldstein's description of the patterns as "vocal gestures" underlines this analogy.

Although the empirical evidence is still incomplete, my proposal is that, by adding the action domain, which is a configurational domain based on the force domain, to conceptual spaces, one obtains the basic tools for analyzing the dynamic properties of actions.

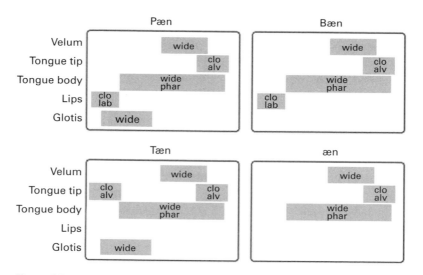

Figure 8.3
Utterances of the four words "pan," "ban," "tan," and "Ann" as sequences of force vectors applied to the five vocal organs. The horizontal axis in each diagram represents the time dimension. Reprinted with permission from http://www.haskins.yale.edu/research/gestural.html.

8.3   Action Categories

To identify the structure of the action domain, we should investigate *similarities* between actions. This can be accomplished by basically the same methods used for investigating similarities between objects. Just as there, the dynamic properties of actions can be judged with respect to similarities: for example, *walking* is more similar to *running* than to *waving*. Very little is known about the geometric structure of the action domain, except for a few recent studies that I will present hereafter. I assume that the notion of betweenness is meaningful in the action domain, allowing me to formulate the following thesis in analogy to the thesis about properties:

Thesis about action concepts: An action concept is represented as a convex region in the action domain.[8]

One may interpret here convexity as the assumption that, given two actions in the region of an action concept, any linear "morph" between those actions will fall under the same concept. One way to support the analogy between the thesis about properties and the thesis about actions is to establish that action concepts share a similar structure with object categories (Hemeren, 2008, p. 25). Indeed, there are strong reasons to believe that actions exhibit many of the *prototype effects* that Rosch (1975) presented for object categories. In a series of experiments, Hemeren (1996, 1997, 2008) showed that action categories show a similar hierarchical structure and have similar typicality effects to object concepts. He demonstrated a strong inverse correlation between judgments of most typical actions and reaction time in a word/action verification task.[9]

Empirical support for the thesis about action concepts regarding body movements can also be found in Giese and Lappe (2002). Using Johansson's (1973) patch-light technique, they started from video recordings of natural actions such as walking, running, limping, and marching. By creating linear combinations of the dot positions in the videos, they then made films that were morphs of the recorded actions. Subjects watched the morphed videos and were asked to classify them as instances of walking, running, limping, or marching, as well as to judge the naturalness of the actions.

The results are presented in figure 8.4. Each of the four prototypical actions is one corner of a pyramid, as shown in the upper right-hand corner of each picture; the cells in between represent the morphed actions as linear combinations of the prototypical actions. The positions of the points in the pyramid symbolize the relative contributions of the four prototypes

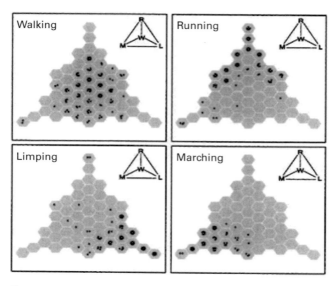

Figure 8.4

Graphical representations of morphs between actions as a pyramid of linear combinations. The letters *W*, *R*, *L*, and *M* symbolize the four prototypes of walking (center), running, limping, and marching. Reprinted from M. A. Giese and M. Lappe. (2002). Measurement of generalization fields for the recognition of biological motion. *Vision Research*, *42*, 1847–1858, © 2002, with permission from Elsevier.

to the morphs. The size of the black dots in each cell represents the number of subjects who judged the morphed action as the action specified.

Giese and Lappe did not explicitly address the question of whether the actions recognized as walking, running, limping, and marching form convex regions in the force domain. However, their data—as presented in figure 8.4—clearly support this thesis. They constructed a variant of the experiment in which the chosen prototypical actions were highly dissimilar: walking, knee bending, aerobics, and boxing. In this case too, the morphed actions were, to a large extent, classified into convex sets.[10]

A third example is Malt et al. (submitted), who studied how subjects named the actions shown in thirty-six video clips of different types of walking, running, and jumping. The subjects were native speakers of English, Spanish, Dutch, and Japanese. The researchers calculated the most commonly produced verb for each clip in each language. This generated a number of verbs also including several subcategories. Another group of subjects, again native speakers of the four languages, judged the physical

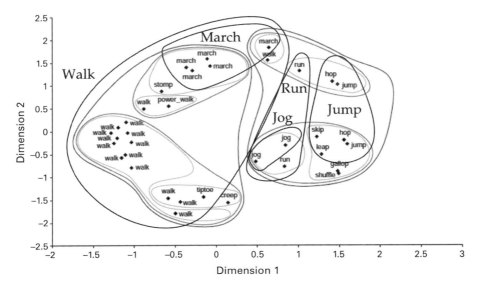

Figure 8.5
Mapping of English names to a multidimensional scaling solution to similarity judgments of video clips. The regions for the main words have been marked (based on Malt et al., submitted, fig. 4).

similarity of the actions in the video clips. Based on these judgments, a two-dimensional multidimensional scaling solution was calculated. The verbs from the first group were then mapped onto this solution. Figure 8.5 shows the results for the most common English action word for each video clip. "March" is a subcategory of "walk," so the region for "walk" extends over the regions for "march." Similarly, "jog" is a subcategory of "run." As can be seen from figure 8.5, the regions corresponding to the names are convex.

Furthermore, the main action words are mainly sorted along dimension 1 of the diagram. Although the multidimensional scaling dimensions do not necessarily correspond to any cognitive domain, a natural interpretation of this dimension is that it represents an increasing amount of force involved in the performance of the action—from walk to run and to jump/hop. Dimension 2 seems more difficult to interpret.[11] Also for the other three languages, the regions for the corresponding action words are convex (see Malt et al., submitted, fig. 4). These results provide additional support for the convexity thesis for actions, though the dimensions of the multidimensional scaling solution do not necessarily directly represent the force domain.

In this section, I have proposed the thesis about actions as an extension of the thesis about properties, for actions that apply to the action domain. The evidence I have presented in support of the thesis about actions is admittedly of an indirect nature. In general, there seems to be a paucity of data on the perception of forces and the ways that forces affect how actions are categorized.[12] More research on action concepts and the postulated force domain is required to establish the validity of the thesis about actions. The claim underlying the thesis is that the same basic mechanism for creating convex regions applies when forming perceptual categories as for action categories.

8.4   Representing Functional Properties in Action Space

Another large class of properties that are difficult to analyze in terms of the object category space in a conceptual space comprises the *functional* properties that are often used for characterizing artifacts. An amusing description of the role of functional properties comes from Paul Auster's book *City of Glass*:

> Not only is an umbrella a thing, it is a thing that performs a function—in other words, expresses the will of man. When you stop to think of it, every object is similar to the umbrella, in that it serves a function. A pencil is for writing, a shoe is for wearing, a car for driving.
>
> Now my question is this. What happens when a thing no longer performs its functions? Is it still the thing, or has it become something else? When you rip the cloth off the umbrella, is the umbrella still an umbrella? You open the spokes, put them over your head, walk out into the rain, and you get drenched. Is it possible to go on calling this object an umbrella? In general, people do. At the very limit, they will say the umbrella is broken. (Auster, 1992, p. 77)

In agreement with Auster's intuition, Vaina (1983) notes that when we decide whether an object is a chair, the perceptual dimensions of the object, like its shape, color, texture, and weight, are largely irrelevant or at least extremely variable.[13] Since I have focused on such domains in my description of object categories, the analysis of functional properties seems to be an enigma for my theory.

I propose to analyze these properties by reducing them to the actions that the objects "afford." To continue with the example, a chair is prototypically an object that affords back-supported *sitting* for one person, that is, an object that contains a stable flat surface at a reasonable height from the ground and another flat surface that supports the back (see also Jackendoff, 1987b, p. 105). The notion of "affordance" is borrowed from

Gibson's (1979) theory of perception. However, he interprets the notion realistically, that is, as independent of the viewer, while for me the affordances are always identified in relation to a conceptual space.

In support of this analysis, Vaina (1983, p. 28) writes: "The requirement for efficient use of objects in actions induces strong constraints on the form of representation. Each object must first be categorized in several ways, governed ultimately by the range of actions in which it can become involved." Following this idea, I propose that function concepts be interpreted with the aid of the *action domain*. To be more precise, I put forward the following special case of the thesis about properties:

Thesis about functional properties: A functional property of an object is a convex set of actions where the object is involved in the underlying force patterns.

The actions involved in the analysis of a functional property may, via the thesis about actions, be reduced to force-dynamic patterns, as I have explained earlier. This is accomplished by representing a functional property as a vector in a high-dimensional space where most dimensions are built up from the force dimensions of the action domain. Functional properties are thus configurational properties in the sense that was described in section 2.5.4. The main problem with this proposal is that we know even less about the geometry and topology of how humans (and animals) structure the action domain than we know about how they structure the shape domain. This is an area where further research is badly needed.

The upshot of the proposal is that even if this road of analysis is long and to a large extent unexplored, functional properties can, in principle, be explained in terms of more basic domains, in particular forces.

Many artifact categories, for example, umbrellas, are identified with the aid of their functional properties or a combination of functional and perceptual properties. Many types of objects have several functions. A sofa, like a chair, affords sitting, but now more than one person can be seated. A sofa also affords lying down. A stool also affords sitting, but without the support for the back. Furthermore, stools and many chairs are movable and afford standing on, while sofas normally do not. These examples show that artifacts can have overlapping functional properties and that some of them have multiple functions.

Whether Auster's object is categorized as an umbrella or not depends on the relative weights that are put on the shape and action domains respectively. The assignment of weights may be highly context dependent. It may depend on cognitive development too: DeLoache, Uttal, and Rosengren

(2004) showed that young children dissociate between actions and perception in their interaction with objects. Two-year-olds attempt to sit on miniature chairs, slide down miniature slides, put dolls' shoes on their feet, and physically force themselves into toy cars: in all cases, the objects are far too small for the action attempted. It seems that the children correctly perceive the action affordances of the object but fail to take the size of the objects into account when trying to perform their actions. In other words, the weight of the size domain used in their planning of action is much smaller than for older children and adults. The dissociation found in younger children provides evidence for the cognitive separation of object category space and the action domain.

In conclusion, the main thesis in this chapter is that actions can be represented by force patterns. The analysis of actions thus highlights the importance of the force domain. In the following chapter, this analysis is embedded in a cognitive model of events.

9 Events

9.1 A Two-Vector Model of Events

With the analysis of properties, object categories, and actions as background, in this chapter I put forward a model of *events* based on conceptual spaces.[1] The model presents events as complex structures that build on other domains, in particular the action domain, object category space, visuospatial domain, and goal domain. I begin with a model of single events and then introduce *event categories*.

When describing events, one must importantly distinguish three different approaches:

(i) *Metaphysical analyses* describing the ontology of events.

One finds several such accounts in philosophy, in the works of Davidson (1967), Kim (1976), Casati and Varzi (2008), and others. However, I agree with Zwarts (2006) that when it comes to the role of events in semantics, these theories are wanting. According to him, the *shape* of the event, that is, "the trajectory or contour that is associated with that event in space or in a scalar or conceptual domain," allows for a more general treatment of semantic phenomena. The model presented in this chapter can be seen as a way of explicating the shape of an event.

(ii) *Cognitive models* of events that account for how humans (and perhaps other animals) represent events mentally.

The model I propose is of this kind. I distinguish between (a) *mental models* of events, which contain representations of causes and effects, and (b) *construals*, which form the semantic basis for utterances. A construal is a mental model of an event with a particular focus of attention (topic) selected from it (see Langacker, 1987, sec. 3.3; Givón, 2001; Croft & Wood, 2000; Langacker, 2008, chap. 3; Croft, 2012a, sec. 1.4).[2]

(iii) Studies of *linguistic expressions* describing construals of events.

In linguistics, a tight mapping between linguistic expressions and construals of events is in general assumed (see, e.g., Rosen, 1999). DeLancey (1991) provides good linguistic arguments why events must be distinguished from their construals. Events are often modeled using symbolic notation (Jackendoff, 1990; Rappaport Hovav & Levin, 1998). For example, Rappaport Hovav and Levin (1998, p. 116) represent the meaning of the verb *break* as follows:

$$[[X \ ACT_{<MANNER>}] \ CAUSE \ [Become \ [Y \ <BROKEN>]]]$$

This can be rendered as "X acts in a manner to cause Y to become broken." In this kind of analysis, one never really leaves the linguistic level, since the verb "break" reappears as <BROKEN>. As a consequence, the approaches (ii) and (iii) are sometimes not clearly separated. In contrast, the model of events in this chapter is constructed from vectorial representations in conceptual spaces. Thus events and construals are clearly separated from linguistic expressions.[3]

With the aid of vectors representing changes of objects, I can define three important notions very naturally:

• A *state* is a set of points in a conceptual space.

Borrowing the notion of state spaces in physics, each object and its properties are identified with a high-dimensional point in a space.

• A *change of state* is represented by a (nonzero) vector.

The beginning of the (high-dimensional) vector represents the initial state, and the end represents the result of the change.

• A *path* is a continuous series of changes of states.

In its original meaning, a path is a series of changes in physical space, but its meaning can naturally be extended to changes in more abstract spaces. The three notions already provide some of the ingredients for an event, which can be constructed quite naturally from the components of conceptual spaces representing the meanings of adjectives (*properties*) and nouns (*concepts*).

A prototypical event is one in which the action of an agent generates a force vector (more generally a force pattern) that affects a patient, causing a change in the state of the patient (more generally a path of changes).[4]

As a simple example, consider the event of Oscar pulling a sledge to the top of the hill (fig. 9.1). In this example, the force vector of the pulling is

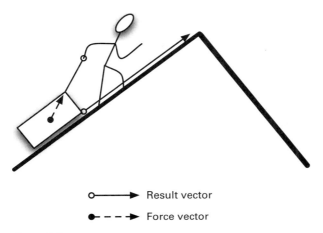

○——▶ Result vector

●— — —▶ Force vector

Figure 9.1
The vectors involved in the event of Oscar pulling a sledge to the top of the hill.

generated by an agent (Oscar). The result vector is a change in the location of the patient—the sledge (and perhaps a change in some other of its properties; e.g., it is getting wet). The result depends on the properties of the patient along with other aspects of the surrounding world: in the depicted event, for example, gravitation and friction act as *counterforces* to the force vector generated by Oscar. (These counterforces explain why the result vector is not parallel with the force vector.)

Another event is the sledge sliding down the hill. Then the force vector is given by gravitation, and the result vector is the movement of the sledge. In this event there is no agent.

As the example shows, an agent is not a necessary component of an event. Although prototypical event representations contain an agent, there are events without agents, for example, events of falling, raining, drowning, dying, and growing. These types of events involve force patterns (gravitation in falling and raining, and internal forces in dying and growing), but no agent generates the forces (unless nature is seen as an agent). Remember that I am focusing on cognitive representation of forces, not what scientific theories say about them.

A central part of the event is also the change of properties in other domains of the patient space. For example, the location of the patient may change—or its color, if the event involves the action of painting. Furthermore, if the force vector is a force pattern, the result may be a path of changes.

Figure 9.2
The main components of a cognitive representation of an event.

Generalizing the examples, I propose the following necessary requirement on the cognitive representation of an event:

The two-vector condition on events: an event must contain at least two
vectors and one object; these vectors are a *result vector* representing
a change in properties of the object and a *force vector* that causes the
change.[5]

The structure of the event is determined by the mapping from force
vector to result vector. I call the central object of an event the *patient*. (In
the linguistic literature, it is commonly called the *theme*.) Just as the force
vector should more generally be described as a force pattern, the result
vector should more generally be described as a sequence of changes, that
is, as a *result path*. For example, if the event is Oscar walking to the library,
his trajectory can be broken down into a sequence of smaller segments
(depending on the desired level of granularity). The model I propose aims
at capturing the central structure of an event (in a spirit not dissimilar from
Dowty, 1991). This means that events can have additional components
not captured here.

A potential problem for the model is raised by Krifka (2012). He argues
that both examples (9.1) and (9.2) hereafter "may be considered true [of
the same event], as simple length measure phrases may refer to paths as
local traces, whereas measure phrases headed by the proposition *for* measure
the spatial extent of the event itself."

(9.1)   Mary walked three kilometers.

(9.2)   Mary walked for six kilometers.

In his opinion, the examples cause problems for an event representation
that has a single result vector. Recall, however, that the model of events is
cognitive, so a mental construal of an event may occur at different levels
of granularity. When Mary walks in a crooked way from the train station
to the city hall, she can be seen as walking three kilometers in a coarse
construal of the event where only the start and end points of the path are

considered. On the other hand, at a finer level of construal of the same event where the different segments of the path are represented as combined result vectors, she can be said to have walked for six kilometers.

The event model captures a basic sense of *causation*: the action of the agent *causes* the change in the patient. The model captures causation by introducing a distinction between forces and changes of states (cf. Wolff, 2007, 2008, 2012). Where there is an effect, there is a cause: if the result vector of an event is nonzero, the force vector must also be nonzero. Equivalently, if the force vector is null, the result vector must be null. This causal condition on event representations is a version of Kant's idea that causation is one of the categories of understanding of human thinking. Wolff and Shepard (2013) go further and argue that there is a direct connection between people's perception of causation and their experiences of touch.

The vectorial representation of forces provides a natural spatialization of causation that unifies the model with other applications of conceptual spaces. In the limiting case when the result vector is the identity vector (with zero length), the event is a *state*. However, an identity result vector does not imply that the force vector is zero. A state can be maintained by balancing forces and counterforces, for example, when a prop prevents a wall from falling.

Using a similar notion of causation, Croft (1991, p. 269) has proposed the following checklist for an "idealized cognitive model of a simple event" as the basis for an analysis of verb semantics:

(i) Simple events are segments of the causal network.

(ii) Simple events involve individuals acting on other individuals (*transmission of force*).

(iii) Transmission of force is asymmetric, with distinct participants as initiator and endpoint.

(iv) Simple events are nonbranching causal chains.

(v) Simple events are independent: that is, they can be isolated from the rest of the causal network.

To a large extent, the model I propose here fulfills these requirements. Regarding (i), I have defined an event as a change, initiated by an agent, that has an endpoint. Events can be chained and are thus segments in a causal network. The focus on agent and patient and on actions analyzed as force patterns fulfills (ii), while (iii) is satisfied because the transmission of force is asymmetric, since the agent initiates the event. Both (iv) and (v) are presumptions of the model. In my opinion, Croft's criteria do not cover all that is relevant for modeling events; his focus on causes and forces misses the way that changes in the properties of the patient are also important for categorizing events.[6]

The model of a prototypical event has some similarities to the image schemas used within cognitive semantics, in particular to the force-dynamic models proposed by Talmy (1988) and Croft (2012a). I compare the model to Croft's in section 9.5.3. The event model presented here is similar to Wolff's (2007, 2008) *dynamics model*. He also includes the force vectors of agent and patient. The main difference is that the model I present embeds his dynamics model within the broader conceptual structure of an event. Since he mainly considers physical movement, he does not model the changes of the patient's properties in a (more general) conceptual space. On the other hand, Wolff considers background forces that I do not include in the basic model (although they show up in some of the event representations).

The ACT-CAUSE-BECOME model of verb semantics presented by Rappaport Hovav and Levin (1998, p. 116) can also be mapped onto the present proposal. The ACT of the formalism corresponds to the force vector, except that not all force vectors involve action. The BECOME corresponds to the result vector (Goddard & Wierzbicka [1994] argue that "happen to" fits better than "become"). CAUSE is the mapping from force vector to result vector. As should be clear by now, the model presented here contains a richer structure provided by conceptual spaces that are grounded in perception and action, and thereby breaks the vicious circle of symbolic representations.

As I will show, more vectors and objects may be involved in event construals. The two-vector model can be seen as a form of basic event schema that can be elaborated by specifying further components. To the minimal representation of an event required by the two-vector condition, a number of other entities (thematic roles) can be added: agent, instrument, recipient, and so on.

A special case of the event model, expressed linguistically by intransitive constructions such as "Susanna is walking" and "Paul is jumping," occurs when the patient is *identical to* the agent. In this case, the agent exerts a force on itself. In other words, the agent modifies its own position in some domain of agent space (= patient space).

In a comment on the event model, Krifka (2012) suggests that temporally situating vectors might considerably enrich the semantics of the event model, in particular to cover certain phenomena related to movement events. The key implication of his suggestion is that vectors do not carry enough information to capture the temporal aspects of events.

I do not deny the role of time in characterizing many events but instead suggest that temporal aspects can be derived more economically from the order of composition of vectors. Since actions and events are dynamic entities—they unfold over time—the time dimension is implicitly assumed in the model.[7] For example, a path implicitly represents time in the order in which changes of states are combined.[8]

The conceptual-spatial structure of the model naturally lends itself to representing the decomposition of events into subevents in at least two ways. First, events can be decomposed into co-occurring or parallel subevents using the dimensions of the patient space. Just as in the real world, the conceptual space within which changes happen can be high-dimensional. I suggest that an event can be decomposed in co-occurring subevents when the result vector expresses changes in multiple domains: if two domains are changed, the change can then be seen as two separate events. For example, if a tire is sliding as well as heating, one may wish to refer to these as separate concurrent events, though they involve the same thing (Bennett, 1996). Such decomposition can be driven by the need to reduce representational and computational complexity to cognitively realistic levels: shifts of attention in construals induce one to focus on different subevents.

Second, events can be segmented sequentially by path subcomponents. As I have shown, a path can be represented as a concatenation of smaller changes, for example, an icicle falling, breaking, and then melting. In this case, the subevents will be a connected subset of change vectors. While this segmentation can correspond to time intervals, it can also be based entirely on the order of changes in the patient space without explicitly introducing the time dimension. In general, verbs do not describe movements in time but describe changes in the visuospatial domain or in the object category domains.[9]

9.2 Agents and Patients

The agent and the patient of an event model are the two most central examples of thematic roles. I model them as objects—albeit sometimes nonmaterial ones—and they can therefore, in general, be represented as points in category space, as analyzed in chapter 6. The domains of the space determine the relevant properties of the agent and the patient.

A patient is an object, animate or inanimate, concrete or abstract. The patient is modeled in a *patient space* that contains the domains needed to account for those of its properties that are relevant to the event that is modeled. Apart from object category space, the properties often include the location of the patient and sometimes its emotional state. A force vector can also be associated with the patient: it represents the (counter) force exerted by the patient in relation to the force vector of the event. This may be a physical force, as when a door does not open when pushed, or an intentionally generated force, as when a person counteracts being pushed. In the representation of events, the patient force vector is often unknown and is taken to be prototypical. This means that the consequences of the force vector of the event are open to various degrees.

An agent is the object—animate or inanimate—that generates the force vector, either directly or indirectly via an instrument. Although I do not provide a full analysis of causation here, suffice it to say that identifying causes with force vectors means that the agent is the one causing something to happen.[10]

An agent is modeled with the aid of an agent space, which minimally contains a *force domain* in which the action performed by the agent can be represented: this is the agency assumption. The force domain is primarily physical, but it can be extended metaphorically to social or mental "forces," for example, commands, threats, persuasions, and seductions.[11] The agent space may also contain the visuospatial domain that assigns the agent a location. In the special case when the patient is identical to the agent—the agent is doing something to itself—the properties of the agent involved in the change must also be modeled.

Dowty (1991) presents what he calls *prototypical agents* and *prototypical patients*.[12] Among his list of properties for an agent proto-role, one finds *volitional* involvement in the event (p. 572). I will treat this as a default assumption about agents: as we shall see, there are also event construals where the agent is nonvolitional, for example, when the agent is a natural force such as a storm breaking a tree.

A stronger assumption about an agent is that it is *intentional*. I conceive of intentionality as involving the agent selecting an action to reach a goal. The goal is represented mentally by the agent, and I model this by a goal domain as part of the relevant agent space.[13]

Empirical evidence from child development research supports this general model of events. First, event representations and the understanding of intentionality develop early in infancy (Nelson, 1996; Wagner & Lakusta, 2009). Michotte's (1963) experiments also show that children assign the roles of agent and patient to moving objects at a very early age. When the agent is animate, children categorize the agent's actions in terms of goals, and not locations or origins (Woodward, 1998). In contrast, no such bias exists for inanimate agents (Wagner & Lakusta, 2009).

9.3   More Thematic Roles

Linguistics has generated a large number of theories of thematic roles.[14] I next want to show that my model is able to accommodate several of the roles that have been proposed. However, I take these roles not as primitive components but as extensions of the basic two-vector model of events and the relations between its parts. The roles are treated as structured positions in a more or less elaborated event schema. In support of this strategy, Levin and Rappaport Hovav (2005, p. 112) write: "What makes event structures appealing is their ability to encode certain properties of events that cannot be represented with semantic role lists, including approaches that treat semantic roles as collections of lexical entailments associated with certain arguments of verbs. . . . For instance, event structures, by having a function-argument form, naturally define hierarchical relations between certain arguments."[15]

I therefore turn to a description of further elements of the model that may be added to account for various aspects of the semantics of verbs. First, one should consider the role of *counterforces*. The force vector of an event construal will change one or more properties of the patient. Elementary operations on vectors provide a reasonable account for how changes can result from compositions of the force vector and the counterforces exerted by the patient. The relevant force vector is the resultant $r = f + c$, where f is the force generated by the action a of the agent, and c is the counterforce of the patient in the original state. In brief, I can define an event as a mapping between an action in an agent space and a resulting change in patient space that results from applying r to the patient.

Second, some actions are *intentional*: the agent exerting the force vector has a goal to make the properties of the patient change in a desired way. As mentioned in section 3.2.6, one way of defining the goal is to assume that the agent has a *value domain*. As I argued in chapter 3, a *goal domain* can be defined as the product of the patient space (representing the possible results of an event) and the value domain of the agent. A value function assigns a point in the value domain to each point in the patient space. Given an initial patient state, an agent will try to reach a point in the patient space that is associated with a maximal value, or at least with a value that is good enough (i.e., *satisficing*, in the terminology of Simon [1969]). For example, if I want to bring my donkey to the stable, I want it to move from its current location to somewhere inside the stable. This can be described mathematically as a choice function from patient space to the action domain such that, for each goal in patient space and each current state of the patient, the function picks out the action that achieves the best (or a satisficing) outcome. In the example, the choice function should, for each possible location of my donkey, specify an action (my pulling the donkey in a particular direction for a sufficient amount of time, overcoming the counterforces of the donkey) so that the resulting change in position of the donkey ends up in the stable.

A consequence of this analysis is that an intentional agent must have a representation of the patient space. Of course, similar actions can be triggered by very different goals: a child hammering on a radiator is aiming for a desirable region of the sound domain, while a plumber performing the same action is aiming for a region of the functional domain of the radiator.

Third, I turn to the "telicity" of actions. Some actions are ongoing in the sense that the force (or better, a force pattern) is exerted for an unbounded amount of time, for example, in walking or pushing an object, with the consequence that there may be no definite endpoint of the result vector (or result path). This is a special case of a more general type of events that are often characterized as *processes* (atelic events). In contrast, there are bounded events (telic events) where the force vector is applied for a limited amount of time and where the result vector (path) has an *endpoint*. An endpoint is a state where the properties of the patient are not changing anymore. An example is when someone reaches a summit; there is no way to go higher. In many languages, the differences between processes and bounded events show up in the syntax. Here I focus on bounded events, but most of the elements of event representations will apply to corresponding representations of unbounded processes. I return to an analysis of aspect in section 9.5.2.

Fourth, many actions involve *instruments*. The typical case is when the agent uses an instrument to exert the force vector, for example, hitting with a hammer or cutting with a knife. Instruments are intermediaries between the agent and the force vector acting on the patient. This can be modeled by breaking down the agency into a chain of vectors. In the special case when the agent is identical with the patient, for example, in walking with a cane, the instrument is used to modify the force pattern the agent applies to herself. In some construals, the linguistic expression of an event focuses on the instrument, for example, "The hammer broke the window." In such a case, the instrument is metonymically made the agent of the event.

Once the thematic roles of agent and patient are represented, it is natural to distinguish between the force vector of the event as applied to the patient and the force vector as generated by the agent. In the second perspective, the force vector typically represents an action. An equivalent force vector applied in pushing an object (from the patient perspective) may be generated by the performance of very different actions (kicking, shoving, leaning, etc.) described by different patterns of forces exerted by the agent. If an instrument is involved, the force exerted by the agent will be modified by the instrument and thus different from the force vector affecting the patient. Hence the force vector of the event should be distinguished from the action of the agent. In any more elaborate description of an event, it is not sufficient to represent only the force and change vectors: the action of the agent must also be included. This will be apparent when I analyze manner verbs in the next chapter.

Fifth, some event construals involve *recipients* in addition to the patient. Common examples are events involving giving and selling. These examples involve intentional actions on the part of the agent. I discuss the role of recipients in section 10.6.

This concludes my presentation of the model of event representations. A general summary is that an event is represented by a number of vectors and a number of entities. The vectors include minimally the force vector and the result vector but may also include the counterforce of the patient, the force vector exerted by the agent, and the intentional goal vector of the agent. The entities include minimally the patient but may also include the agent, an instrument, and a recipient. This list accounts for many of the thematic roles that have been treated in the linguistic literature.

So far, I have considered only single events. The model is easily extended to the chaining of events. The notion of function composition carries immediately the mathematical model of how to chain events. Of course, not all events can be chained; what was the range of the first event must become a part of the domain of the second.

9.4 Event Categories

In general, events should be represented not only as single instances in time and space but more generally as event categories, for example,

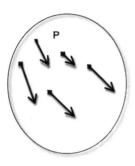

Figure 9.3
Changes represented as vectors in patient space *P*.

climbing a mountain. In this section, I provide a framework for the analysis of event categories, of which single events can be considered instances.

The description of result vectors can be generalized to that of *result vector fields* by associating to each point in patient space *P* the vector change induced by a particular action, taking into account if necessary the (counter) force exerted by the patient. Notice that the vector field represents all the changes induced by applying an action to every conceivable state in patient space *P* (fig. 9.3). This implies that, by (continuously) composing vector changes, one can reconstruct the trajectory (path) in *P* generated by repeated application of an action.

I now have the tools to define an *event category*: an event category is a structure that represents how the agent space affects the patient vector field. Thus an event category should represent not only the mapping from one action to the patient vector field but also the way all actions in the action category of the agent space affect the patient. For example, an event category of *pushing a table* should be able to represent the effect of different, albeit similar, patterns of force on the different points in the table patient space.

The model allows me to represent events at different levels of generality. There are subcategories of events just as there are for object categories. For example, *pushing a door open* is a subcategory of *pushing a door*, where the force vector exceeds the counterforce of the patient. *Pushing a door but failing to open it* is another subcategory, where the counterforce annihilates the force vector.

Can events be represented as convex regions of some suitable space, in the same sense as I have argued for properties, concepts, and actions? At present, I do not have a general answer to this question. A simple mathematical model of an event category is as a region of the space consisting of the product of action space A and the vector field of changes $P \times P$.[16] Although actions are convex regions of A, and properties are convex regions of P, this does not guarantee that events will be convex regions of $A \times P \times P$. The conditions under which this will hold remain to be determined.

The models of actions and events that I have presented in this book are motivated by using conceptual spaces as a framework. What is new—apart from using conceptual spaces—in this model of events and event categories is the introduction of the two vectors associated with force and result. A corresponding model for processes can be developed, but I will not pursue that topic here.

9.5   Three Conceptualizations of Events in Linguistics

My approach has clear connections to the way linguists have conceptualized events to understand verb semantics. Among linguistic approaches, three have gained prominence (Levin & Rappaport Hovav, 2005, chap. 4): the localist, the aspectual, and the causal. The *localist* approach focuses on motion and location, in physical as well as in abstract spaces. The *aspectual* approach puts the temporal properties of events in the center. The *causal* approach highlights the role of causal chains and transmissions of force. These three approaches have been connected with the role of verbs. I want to show that all three approaches can be subsumed under the general model of events presented in this chapter. The proposed model brings out the strengths and limitations of the approaches while providing a unifying common ground for their central features.

9.5.1   Localist Approach
Jackendoff's (1976, 1983, 1990) localist hypothesis claims that all verbs are construable as verbs of motion and location (Levin & Rappaport Hovav, 2005, p. 80). Clearly conceptual spaces provide a suitable background for locating entities: motion is represented by the result vector. In many cases the motion takes place not in physical space but in abstract spaces. As a consequence, Jackendoff (1990) extends the strict localist approach to distinguish change-of-location verbs from change-of-state verbs, two notions derivable as specifications of result vectors.[17] Even including this distinction, however, the localist approach cannot handle verbs that

express the force vectors in a natural way (Rappaport Hovav & Levin, 2002). In localist analyses, an agent is often treated as a "source," but this does not suffice to describe the role of force vectors. I see this as a limitation of the localist approach.

9.5.2 Aspectual Approach

A long tradition within linguistics classifies verbs into aspectual classes by means of the internal temporal properties of the events they express. A classic proposal is that of Vendler (1957), who distinguishes between states, activities, achievements, and accomplishments. He uses three contrastive distinctions: stative versus dynamic, durative versus instantaneous, and telic versus atelic. Using these distinctions, we can describe the four classes as follows:

• A *state* is stative, durative, and atelic. Example: be in the house.
• An *accomplishment* is dynamic, durative, and telic. Example: build a bridge.
• An *achievement* is dynamic, instantaneous, and telic. Example: realize an error.
• An *activity* is dynamic, durative, and atelic. Example: walk.

To Vendler's four aspectual classes, many researchers add a fifth: *semelfactives*—such as "jump," "knock," or "beep"—that are dynamic, instantaneous, and atelic. Originally Vendler intended his classification to apply to verbs, but aspectual classification really involves event descriptions (Jackendoff, 1991, sec. 8.3; Levin & Rappaport Hovav, 2005, p. 90).

In the model presented here, the different aspects can be accounted for by describing the properties of the vectors or paths involved in an event. One distinction to be made is whether the vector is *extended* or *punctual*. An extended path can be decomposed into a sequential composition of subpaths. A punctual one cannot; in other words, you can only consider one moment. To some extent, this distinction corresponds to the durative-instantaneous distinction, except that the time dimension is not represented explicitly in the basic model, for which purpose the decomposition properties of the vectors or the paths suffice. Another distinction concerns whether the vector or path has an endpoint (fixpoint) or not. This corresponds to the telic–atelic distinction. Jackendoff (1991) makes essentially the same distinctions, but he writes about directional dimensions instead of vectors (see also Jackendoff, 1987c).

Given these two distinctions, the proposed theory generates the following classification. In the special case when the relevant vector is

Table 9.1
Classification of aspects

Vector or path	*is extended*	*is punctual*
has fixpoint	accomplishment (build a bridge)	achievement (realize your error)
has no fixpoint	activity (walk)	semelfactive (knock)

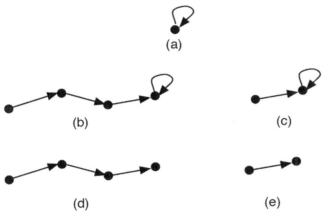

Figure 9.4
Aspects as path properties: (a) state, (b) accomplishment, (c) achievement, (d) activity, (e) semelfactive.

the (zero-length) identity vector, the event is a state. The other four cases are illustrated in table 9.1.

This classification scheme can also be expressed figuratively in terms of vectors (fig. 9.4):[18]

(a) A state is a fixpoint: absence of change.
(b) An accomplishment is a series of subevents leading to a final state (telic).
(c) An achievement is like an accomplishment (telic) but consists of only one instant (no temporal structure can be derived from it, beyond a distinction of before and after).
(d) An activity is like an accomplishment, except that it does not end in a final state.
(e) A semelfactive is an activity happening in one step.

I thus view aspect as primarily functional, not temporal. I distinguish between the representation of an event and the way it is embedded in the time dimension. My analysis is quite similar to the one proposed by Croft (2012a, sec. 2.3), but it is less detailed. For example, Croft includes a temporal dimension, which I do not require.[19] That said, by focusing on path properties, I believe I obtain an even more coherent analysis.

9.5.3   Causal Approach

Nevertheless the model that comes closest to mine is Croft's (2012a, 2012b), specifically his "three-dimensional representation of causal and aspectual structure in events."[20] In his earlier work (Croft, 1991, 1994), he presented a causal model with four basic elements: *initiator, cause, change,* and *state.* The initiator corresponds to the agent in my model, exerting the force vector. His cause is my force vector. His change corresponds to the result vector, and his state to the endpoint of the result vector. His later work, however (particularly Croft, 2012a), presents a geometric model that shares many features with the one presented here. His model is best illustrated by the example in figure 9.5.

This figure represents two subevents. The lower scheme involves two dimensions, *q* (force) and *t* (time), where Jack's action is represented as a momentary change in the *q*-dimension. The upper scheme also involves two dimensions: *q* (qualitative change) and *t* (time), where the change of the vase is represented by a change of level along the *q*-dimension. The arrow from the lower to the upper scheme represents the causal chain. Croft (2012a, p. 9) emphasizes that his model is based on geometrically (as opposed to diagrammatically or symbolically) represented components.

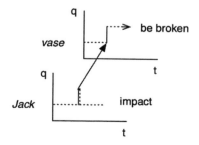

Figure 9.5
Croft's representation of "Jack broke the vase" (from Croft, 2012a, p. 212). Reprinted by permission from Oxford University Press.

According to Croft (2012a, pp. 216–217), his model demonstrates that "events can be decomposed in three distinct ways: temporally, in terms of the temporal phases; qualitatively, in terms of the states defined on the qualitative dimensions for each participant's subevent; and causally, in terms of the segments of the causal chain." He points out that dividing an event into subevents accounts for its causal structure, while the nature of the qualitative changes covers the aspectual structure. In these respects— and furthermore in using a dimensional analysis—his model corresponds to the one proposed here.

Despite the obvious similarities between the two models, certain differences are worth mentioning. Most important, the model presented here focuses more on the *geometric* structure of the domains. Croft writes in general about "qualitative" dimensions, but I include their different geometric and topological structure. Croft treats all qualitative dimensions on a par. What he calls the dimension of qualitative state corresponds to what I call domains. In general, I allow domains to be multidimensional and to have a rich structure.

I should stress that the vectors of the event model do not occur in a vacuum but are always construed in relation to a domain and thus inherit all the structural domain information. For example, because the domain of temperature is one-dimensional, the heating vector has only two directions (up and down), while in the three-dimensional domain of color, multiple directions are possible. In brief, behind the vector is a wealth of information coming from the domains that add semantic content. This argument can be seen as a partial solution to what Chomsky (1986) calls Plato's problem, that is, the fact that we know more about the meaning of almost any word than what we could have learned from our exposures to it.

Hence conceptual spaces provide a more explicit semantic framework for expressing Croft's notion of *quality of states*. In principle, his model can handle the similarities between actions and results—and thereby also the similarities between verbs—but he does not develop this. Another difference from my approach is that Croft does not present a theory of actions to underpin the *force* aspect of the event. The division between the force and result vectors can be seen as implicit in his model, but he never formulates anything like the single-domain thesis for verbs (see sec. 10.3). Furthermore, I require no explicit representation of the time dimension, building everything into the dynamics of the force and result vectors.[21] Croft (2012a, p. 219) acknowledges that his model cannot explain these differences, since "it only distinguishes causal and non-causal relations on the causal dimension." Together these differences provide the model presented here with a richer potential for new predictions concerning the semantics of verbs, as will be seen in the following chapter.

Another causal theory is that of Talmy (1972, 1976, 2000), who distinguishes between four types of causation: volitional (a volitional entity acts intentionally on a physical object), physical (a physical object acts on another physical object), affective (a physical object acts on a sentient entity), and inducive (a volitional agent acts intentionally on a sentient entity, changing its mental state). The components of the proposed model of events can account for all four types.

This concludes my account of how the three perspectives discussed by Levin and Rappaport Hovav (2005) are subsumed by the two-vector model of event.

9.6   From Event Representations and Construals to Sentences

No verbal symbols can do justice to the fullness and richness of thought.
—John Dewey

This part of the book focuses on lexical semantics. In contrast, semantic theories within philosophy mainly concern the meaning of sentences. Everyone in linguistics and philosophy seems to take the status of sentences as a fundamental unit for granted.

The traditional account within analytic philosophy is that a (declarative) sentence expresses a *proposition*. Propositions are taken to be either true or false, that is, to have truth-values. In many accounts, a proposition is identified with a set of *possible worlds*. As I explain in Gärdenfors (2000, sec. 3.3), from the point of view of the six requirements in chapter 1, this is putting the cart before the horse, since possible worlds are cognitively inaccessible entities. Furthermore, most of the examples of sentences in the philosophical discussion involve stative situations; the classical example is "snow is white." In contrast to most philosophical theories, I do *not* assume that a semantic mapping exists between sentences and propositions. The reason is that the meaning of a sentence to a large extent depends on its context. For example, an ironic communicative act may drastically change the standard meaning of a sentence.

Alas, it would lead me into a jungle with a morass trying to account for the drawbacks of the semantics of sentences within analytic philosophy. Instead I want to be more constructive and present the bare outlines of how sentences can be analyzed on the basis of conceptual spaces. If one takes a cognitive-communicative point of view, as presented in the first part of the book, it is not so obvious why we express ourselves in sentences. Frege's answer that the meanings of sentences are *thoughts* is simply not sufficient, since no one knows how a thought is identified (independently of language). So what do sentences mean? In chapters 6 and 7, I have presented analyses of nouns and adjectives, and the following chapter is devoted to the semantics of verbs. What happens to meanings if these parts are put together?

My basic idea is that *sentences express events*. Furthermore, from a communicative point of view, the focus should be on *utterances* rather than on sentences. Utterances are parts of a communicative context that contributes to the meaning, while in the philosophical (and much of the linguistic) discussion, sentences are often analyzed as having a meaning that is independent of the context. However, so as not to clutter the notation, I use the word "sentence" to take in the role of "utterance" as well.

Any description of an event is based on a construal. When I introduced the notion of a construal, I focused on the role of *attention* as a selection mechanism for a cognitive event representation. There are, however, other aspects of how a construal is formed (for a survey, see Croft & Wood, 2000, chap. 3; Langacker, 2008, chap. 3). One aspect is *perspective*: for example, if you and I are located on two sides of a house, I can say that you are behind the house, if I put myself in the center, or I can say that you are in front of the house, if I put the house and the direction of its main side in focus. Another aspect is *categorization*: a construal must select a level of generality to describe an object, for example, *terrier*, *dog*, *mammal*, or *animal*. Yet another aspect is the relation to the *common ground* in the communicative situation. For example, when selecting whether to use a pronoun, noun, or name to refer to an individual, the speaker must consider whether the individual or the name is part of the common ground.

What minimal elements must a construal of an event contain? A generic way of describing an event is that "something happens to something." According to the model of events presented earlier, the something it happens to is the *patient* (sometimes also the agent). Furthermore "happen" is a placeholder for either the force or the result vector. This leads to the following thesis:

Thesis about construals: A construal of an event contains at least one vector (force or result) and one object.

On the basis of the notion of a construal of an event, I can now formulate a fundamental connection between the semantics of sentences (utterances) and events:

Thesis about sentences: A (declarative) sentence typically expresses a construal of an event.

From a communicative perspective, one can ask why sentences have such a fundamental status in comparison to other compositions of words. I do not believe that the question has a unique answer, but I will base my analysis on the levels of communication discussed in section 5.2. On level 1, instruction, sentences are often not required. If one is sitting at a dinner table, "Salt!" may function as request, albeit not a polite one (polite requests are often concealed as questions). Or if one is standing in front of a door, "Open!" may be an efficient speech act, since the addressee (the agent) and the object are contextually given. When it comes to level 2, coordination of inner worlds, the situation is different. Here the communication typically concerns agents, patients, actions, and results that are not present in the context of the utterance. Then the thesis about sentences implies that at least an agent or a patient (expressed by a noun phrase) and a force vector or a result vector (parts of a construal of an event) are parts of what is expressed. Thus the two main components, noun phrase and verb phrase, have to be present in a linguistic description of an event. The upshot is that on the coordination level of communication, sentences are indeed central units. On the third level, coordination of meanings, generics are typically used. Generics are sentences, but atypical ones, since they describe generic information about concepts rather than events.

For these reasons, sentences are natural units of a semantic theory, albeit not as central as philosophers and some linguists want. In brief, the model of events and the thesis about construals explain the central components of a sentence. They thus provide a motivation for a sentence being a *cognitive unit* of communication.[22]

In this chapter, I have shown that a model of an event can be a complex structure, involving not only the two vectors, a patient and an agent with their properties, but also counterforces, instruments, recipients, intentions, and so on. Although the mental model of an event may be complex, a sentence (utterance) captures only certain features of a construal generated from a particular focus on the event.[23] By analogy with the visual process—where we can only focus our attention on some features of the visual field—a construal focuses only on certain parts of an event. The sentences "Victoria hits Oscar" and "Oscar is hit by Victoria" describe the same event with the aid of two different construals, where Victoria and Oscar, respectively, are put in focus.[24] Another example is that the difference between the sentences "Pat sprayed paint on the wall" and "Pat sprayed the wall with paint" is that in the first sentence, "paint" is focused on as the patient of the action, while in the second, "the wall" is made the patient (Levin & Rappaport Hovav, 2005, p. 124).

Consequently no simple mapping exists between the role taken in an event and the designation of subject, object, or oblique. A sentence expresses a construal representing a particular focus on an event. In English (and many other languages), the most focused role is designated subject, and the secondary focus is designated object. Givón (2001) calls these *primary* and *secondary topics*. He writes that topicality "is fundamentally a cognitive dimension, having to do with the focus on one or two important event-or-state participants during the processing of multi-participant clauses" (p. 198). As Croft (2012a, pp. 252–253) notes, this phenomenon creates problems for all argument realization rules that are based on thematic roles.[25] In agreement with Givón (2001, p. 198), I see topicality not as directly part of event representation but as a central element of the construal process. This setup avoids the problems that arise when event representation and construal are conflated. Speakers have conversational goals in producing construals. Consequently the construals are contextual, depending on what the conversation partner already knows or believes or will find most interesting.

10   The Semantics of Verbs

In this chapter, I apply the model of events and construals to show that it can form the basis for a general semantics of verbs. In linguistics, the analysis often starts with a particular syntactic feature, and then one tries to find what is semantically common to what is expressed by this structure. For example, Levin and Rappaport Hovav (2005, p. 131) write that their work "is predicated on the assumption that there is a relationship of general predictability between the lexical semantic representation of a verb and the syntactic realization of its arguments."[1] It should be clear, though, that no unique path leads from event construals to linguistic realization; different solutions are found in different contexts and in different languages. Nevertheless the underlying semantics *constrain* the syntactic structures.

10.1   Verbs Refer to Vectors in Events

In linguistics, the semantic role of verbs has been described as *predication* (e.g., Croft, 2001). However, the notion of predication is rather abstract (it derives from the Fregean view of language), and it does not describe the communicative role of verbs.[2] Furthermore, adjectives are used predicatively too, so the notion does not characterize the use of verbs.

Following my general strategy, I want to identify the communicative functions of verbs before formulating the central thesis concerning the semantics of verbs. Again I rely on the three levels from section 5.2. On level 1, instruction, the verb obviously has an extremely central function: it is difficult to give a linguistic instruction without using a verb. From an evolutionary point of view, it is likely that imperatives involving something with the semantic function of a verb were among the first constructions in a protolanguage.

The use of *commands* now becomes particularly relevant. In chapter 4, I presented the three protomodes of communication: imperative pointing (protoimperative), emotive and information-requesting declarative pointing (protointerrogative), and goal-directed declarative pointing (protodeclarative). As shown there, imperative pointing and goal-directed declarative pointing can help in achieving basic forms of coordination of action.

When it comes to commanding actions, however, pointing is not sufficient. This is where verbs prove their mettle. One type of command is directed toward achieving a particular result: "Open the door," "Tom, paint the fence," or "Dress the king." Another type is directed toward the type of action required: "Rake the leaves," "Dig a ditch," or "Run as fast as you can to the doctor." I see these uses of verbs as their most basic communicative function (cf. Wittgenstein, 1953).[3]

On level 2, coordination of inner worlds, verbs are necessary components in linguistic descriptions of events according to the thesis about construals. Verbs have two main roles: (i) to describe *what* has happened (or will happen), and (ii) to describe *how* it happened (or will happen). Cooperative planning may require a substantial coordination of what has happened with the planned sequence of results that are to be achieved by the cooperating partners.[4] An extension of this form of coordinating inner worlds is *narration*. A narrative is a description of a causally related sequence of events—actual or imagined.

On the third level, coordination of meaning, verbs have a role in generics that seems to be of the same nature as that of nouns and adjectives.

My analysis begins from construals of events, and I aim to identify *lexicalization constraints* for verbs. I focus on the meanings of verb roots, since the variety of possible syntactic modifications will make a full semantic analysis of verbs very complicated. The key idea for the semantics of verbs is the following:

Thesis about verbs: A verb root refers *either* to the force vector *or* to the result vector of an event (but not both).

The main importance of this thesis is that it connects two of the central components of the model of events directly to a particular class of words.

10.2 Similarity of Verb Meanings

Before presenting further lexicalization constraints for verbs, I want to point to some predictions from the thesis about verbs. First of all, the vector representation explains *similarities* of verb meanings, by building on the distances between underlying action vectors. That the meaning of "walk" is more similar to that of "jog" than that of "jump" can be explained by the fact that the force patterns representing walking are more similar to those for jogging than those for jumping. Although I have not presented the details of the similarities of the actions involved, they can be worked out systematically from the vectorial representation of actions as indicated in chapter 8.

In a parallel way, the thesis explains the general pattern of the *subcategorizations* of verbs by exploiting categorizations of actions: For example, the force patterns corresponding to the verbs "march," "stride," "strut," "saunter," "tread," "limp," and so on, can all be seen as subsets (more precisely, subregions) of the force patterns that describe "walk." The inference from, for example, "Oscar is marching" to "Oscar is walking" follows immediately from this inclusion of regions within one another. Notice that such an inference cannot be explained if the meanings of verbs are expressed symbolically, as in the ACT-BECOME example in section 9.1. As far as I have found, no previous theory of verb semantics can account for these two central properties.

The analysis extends to metaphorical uses of verbs. I claim that an important type of metaphor is based on similarities of force vectors in line with Lakoff's (1990) "invariance hypothesis." For example, when a football player "scythes down" another player, the metaphor builds on the similarity between the force patterns involved in scything crops and the movement of the first player's legs in relation to those of the second player. Another metaphor of the same type is when a tennis player "slices" a backhand. Other examples are when an employee "pitches" a proposal to the boss, where physical action from throwing in baseball is transferred to an action in the mental domain concerning relations between people, and when someone "breaks into" a conversation, where physical action is transferred to an event occurring in the time domain.

Finding the force invariances involved in such metaphors seems to require a fairly advanced form of abstract thinking. For example, Seston, Michnick Golinkoff, Weyiy, and Hirsh-Pasek (2009) show that eight-year-old children, but not six-year-olds, can understand sentences such as "When Taylor spilled his milk on the table, he vacuumed it up with his mouth" as well as adults do. The force vectors involved in vacuuming are sufficiently similar to Taylor's action with his mouth that the older children can map them onto the situation described in the sentence.

10.3   The Single-Domain Thesis for Verbs

10.3.1   A Central Thesis for Verbs

Verbs cannot mean just anything. Kiparsky (1997) suggested that a verb expresses inherently at most one semantic role, such as theme, instrument, direction, manner, or path. Rappaport Hovav and Levin (2010, p. 25) strengthened this idea by associating semantic roles with argument and modifier positions in an event schema and proposed that "a root can only be associated with one primitive predicate in an event schema, as either an argument or a modifier."

By grounding meanings not in a symbolic event schema (as, for example, do Rappaport Hovav & Levin, 1998, p. 109) but in conceptual spaces, I can, by using the notion of a domain, refine and strengthen the constraints proposed by Kiparsky and by Rappaport Hovav and Levin (see also Kaufmann, 1995):

The single-domain thesis for verbs: The meaning of a verb root is a convex region of vectors that depends only on a single domain.

For example, "push" refers to the force vector of an event (and thus the force domain), "move" refers to changes in the visuospatial domain of the result vector, and "heat" refers to changes in the temperature domain.[5]

The single-domain thesis for verbs is analogous to the thesis that adjectives denote convex regions in single domains that I presented in chapter 7; that is, there are no adjectives that mean, for example, "red and tall" (multiple domains), and there are no adjectives that mean "red or green" (not convex). Likewise there are no verbs that mean "walk and burn" (multiple domains), and there are no verbs that mean "crawl or run" (not convex).[6]

Regarding the convexity of verb meanings, the verbs referring to force vectors or force patterns for actions derive their convexity from the corresponding actions (see sec. 8.3). The convexity for result vectors based on a singular dimension is easy to prove. For example, consider "heat." The result vector $<t_1,t_2>$ consists of an initial temperature t_1 and a resulting temperature t_2, where $t_2 > t_1$. For any two vectors $<t_1,t_2>$ and $<t_3,t_4>$ where $t_2 > t_1$ and $t_4 > t_3$, and for any vector $<t_5,t_6>$ where t_5 is between t_1 and t_3 and where t_6 is between t_2 and t_4, it holds that $t_6 > t_5$, which proves convexity. This argument can be extended to domains involving several dimensions.

The single-domain thesis can be seen as strengthening the thesis about verbs: if a verb only refers to a region of vectors in a single domain, then it cannot refer to both the force and the result vectors of an event. Since the model presented in the previous chapter requires that an event always contains two vectors, the single-domain thesis also entails that a single verb cannot completely describe an event but only bring out an aspect of it. However, the two-vector constraint has the testable consequence that a construal can always be expanded to contain references to both the force and result vectors. More precisely, for any statement based on a construal involving only a force vector, one can always meaningfully ask, "What happened?" (filling in the result vector); and for any statement based on a construal involving only a result vector, one can always ask, "How did it come about?" (filling in the force vector).[7]

The result vector of an event represents the change in the properties of the patient. In general, that change can involve multiple domains. For example, when a plant is growing, it not only changes its size but will change shape and possibly color as well. However, the single-domain thesis requires that construals of events only concern changes in one domain. In other words, the focus of attention is on one aspect of the event only.

A fundamental question is: how can the single-domain thesis be cognitively motivated? Why are there no verbs that refer to more than one domain, for example, verbs that cover both the force and result vectors? Kemp and Regier (2012, p. 1049) present a general principle of cognitive economy with respect to semantics: "Categories tend to be simple, which minimizes cognitive load, and to be informative, which maximizes cognitive efficiency." A reason for minimizing cognitive load that was proposed in Warglien, Gärdenfors, and Westera (2012) builds on *learnability* constraints: each domain contains an integral set of dimensions that is separable from other domains. A verb involving a mapping between domains may be hard to learn and subject to many contingencies and sources of instability. For example, a change in location of a fruit and a change of its taste are not correlated. No corresponding domain combines these domains: consequently no verb exists that simultaneously expresses change in location and change of taste.

In particular, the coupling of force and change vectors is complicated, since it concerns the way actions relate to their effects. For example, one understands well how patterns of forces exerted by one's arms lead to different actions, for example, when hitting a tennis ball; the movement of a physical object (tennis ball) is likewise well understood; but the relationship between the two is unstable, being subject to unknown counterforces and other uncontrollable factors. It is therefore difficult to learn.

Admittedly, the strength of the thesis depends partly on how domains are identified just as for adjectives. As has been discussed earlier, for some meaning areas, it may be problematic to identify the appropriate domain. For example, it may seem difficult to reconcile verbs involving social relations like *partying* with a single domain. I see it as a research program to analyze the domains presumed by different verbs to test the viability of the single-domain thesis.

10.3.2   Putative Counterexamples

An immediate consequence of the single-domain thesis is that *no verb can express both the force domain and another type of domain.* The literature contains several putative counterexamples, for example, "climb" (Jackendoff, 1985; Goldberg, 2010; Kiparsky, 1997; Levin & Rappaport Hovav, forthcoming).

(10.1)   Oscar climbed the mountain.

(10.2)   Oscar climbed down the mountain.

(10.3)   Oscar climbed along the rope.

It seems that, in its prototypical sense (10.1), climbing involves both upward motion (result) and clambering (manner), while in other uses (10.2, 10.3) the motion has another direction. However, the single-domain thesis is fulfilled by noting that the force vector of "climb" is required to have an upward direction (cf. Geuder & Weisgerber, 2008; Rappaport Hovav & Levin, 2010). This constraint on the force vector typically generates an upward motion (the result vector), but as (10.2) and (10.3) show, exceptions can be made, marked by a preposition describing the direction of the result vector.

(10.4)   The train climbed the mountain.

(10.5)   ?The train climbed down the mountain.

In (10.4) the force exerted by the train still has an upward direction (though very slanted), but it is only metaphorically a case of clambering. However, in (10.5) the force exerted by the train no longer has an upward direction, and so "climb" is less successfully applied in events of this type.[8] The examples all indicate that the upward direction of the force vector is a central ingredient of the meaning of "climb."[9]

(10.6)   The snail climbed up the side of the tank.

Levin and Rappaport Hovav (forthcoming) consider examples like (10.6) to be counterexamples to the requirement of clambering as part of the meaning of "climb." However, the snail's use of suction should be seen as a metaphorical form of clambering: the force patterns involved are sufficiently similar.

Kracht and Klein (2012) argue that verbs of (mental) coercion ("persuade," "force," "compel," etc.) violate the single-domain constraint. They claim that these verbs lexicalize both a manner and a result component. In contrast, I do not view these verbs as counterexamples to the single-domain constraint. The reason is that constructions with these verbs involve *another* verb. The form is "X persuades/forces/compels Y to V," where V is a verb. The coercion verb is a manner verb in an event generated by an agent X acting on a patient Y that combines with another verb describing another event where Y is agent.[10] The second event can be seen as a result of the coercion event, but this does not mean that the coercion verb lexicalizes a result component. The verb predicts that a result took place, not the content of the coercion. This is the job of the second verb V. As a matter of fact, V can be almost any verb—manner or result. What makes the coercion verbs special is that the result of the force vector is not a result vector but a new event. In brief, coercion verbs have a more complicated meaning structure than many other verbs, but they do not invalidate the single-domain constraint.

Wolff (2012) also argues that verbs of causation—for example "cause," "allow," "enable," "block," "prevent"—cannot be single domain, since they involve both force and result. However, his examples involve interactions between two force vectors (from the agent and the patient), a case that is not covered by my basic model, but it does not really leave the single domain of forces. As with the coercion verbs, these cases always require either another verb, for example, "X prevents Y from V." Again, the causation verbs concern more complicated situations, involving one event with an agent and a patient and another event where the patient of the first is agent.

10.4 Manner and Result Verbs

10.4.1 Manner/Result Complementarity
Traditionally (Talmy, 1975, 1985; Levin & Rappaport Hovav, 1991) there have been two main ways of dividing verbs into classes:

(i) manner versus path, as in "jog" versus "cross"
(ii) manner versus result, as in "wipe" versus "clean"

A direct consequence of the single-domain thesis is that the distinction between the different kinds of verbs is determined by the domain associated with a verb. If the domain is that of the force patterns underlying actions, it is a manner verb. If it is the visuospatial domain, it is a path verb. For all other domains, for example, object category domains, it is a proper result verb. Thus the single-domain thesis together with the classification of domains can explain why these three kinds of verbs fall out as natural classes.

Levin and Rappaport Hovav (Levin & Rappaport Hovav, forthcoming; Rappaport Hovav and Levin, 2010) simplify the two divisions to just one by distinguishing between *manner verbs* and *result verbs*—where "manner verbs specify as part of their meaning a manner of carrying out an action, while result verbs specify the coming about of a result state" (Rappaport Hovav & Levin, 2010, p. 21; see also Talmy, 2001, chap. 1). Rappaport Hovav and Levin (2010, p. 22) claim that any verb "tends to be classified as a manner verb or as a result verb." Path verbs can be grouped together with verbs that describe property changes because of the tendency to give the same linguistic construction to a changing entity as to a moving one (Gruber, 1967; Jackendoff, 1972; Pinker, 1989, p. 47): both involve changes of properties of the patient, which manner verbs do not.

A consequence of the thesis about verbs is that manner verbs refer to force vectors of events, while result verbs refer to result vectors. Another way of expressing this is to say that the manner/result distinction is basically a cause/effect distinction: manner verbs refer to causes, and result verbs to effects.[11]

Rappaport Hovav and Levin (2010) derive their thesis from their ACT-BECOME model of events together with the constraint that a verb root can only be associated with one primitive predicate in an event schema, as either an argument or a modifier. Since they assume that manner roots modify the predicate ACT and result roots are arguments of BECOME, the manner/result complementarity follows (Rappaport Hovav & Levin, 2010, sec. 2). As noted in the previous chapter, the distinction between force and result vectors corresponds quite clearly to their ACT and BECOME, but it adds a grounding in conceptual spaces that allows more predictions.

10.4.2 Putative Counterexamples

Rappaport Hovav and Levin's version claims that the manner/result complementarity is inherent in the lexical meaning of a verb root. This claim has been criticized (Goldberg, 2010; Koontz-Garboden & Beavers, 2012). In particular, there are verbs that can seemingly be used to express both manner and result, for example, verbs of killing (Koontz-Garboden & Beavers, 2012) such as "drown," "hang," and "crucify" and verbs of cooking (Goldberg, 2010) such as "roast," "fry," and "stew." Furthermore, Goldberg (2010, p. 48) discusses verbs of creation, in particular cooking verbs, that seem to involve both manner and result: "The difference between *sauté*, *roast*, *fry* and *stew* would seem to involve the manner of cooking and yet there is arguably a directed change as well, as the concoction becomes sautéed, fried or stewed." In my opinion, this is only an example of a very strong expectation of the result of the action. Still, when the verb occurs together with an agent, it is an intentional manner verb.[12] What complicates the situation is that unaccusative (anticausative) uses of these verbs also exist, for example, "The fish is frying" and "The pork is roasting"— where the verb is a result verb. In the unaccusative case, the intentional component of the meaning is absent. These verbs thus have a double use. In each instance, however, they will be either a manner verb or a result verb.

Therefore, I propose instead to identify a basic manner or result use of a single verb, and offer the following modified version of Rappaport Hovav and Levin's (2010) constraint:

Thesis about manner/result complementarity: Each verb root has a default manner or result type of meaning, but the other type can also become lexicalized.

This version of the complementarity still claims that a manner or result component is the default meaning, but it allows that some verbs have a secondary meaning that can override the default in specific contexts.[13] The proposed thesis is still compatible with the single-domain constraint.

Two common mechanisms for meaning extension should be mentioned. The first is that a manner verb root such as "fry" is transformed in a metonymic way to the strongly expected result that the food becomes fried. The result meaning of the verb is then sometimes used unaccusatively: "The chicken is frying."

The second mechanism operates when a result verb such as "sink" is transformed into a manner verb with the meaning "cause to sink" by reprofiling the event as in, for example, "The bomber planes sank the destroyer." I realize that to make this story complete, I need to present a way to test which meaning of a verb is primary. Of course, which verbs have a second meaning lexicalized depends on the history of a particular language and varies between languages. However, when the link between manner and result is reliable, we should see a tendency for either of the two mechanisms to apply, and for a verb to gain a secondary meaning.

These two semantic mechanisms are special cases of what Traugott (Traugott & Dasher, 2002, pp. 34–41; Traugott, 2012) calls "invited inferences."[14] The invited inferences are expectations generated from the meaning of a word. The expectations can change over time as a result of new uses, for example, introduced by a metaphor, a metonymy, or new

Table 10.1
Stages in semantic change (based on Enfield, 2003, p. 29)

	Stage 1	Stage 2	Stage 3	Stage 4
Form	w	w	w	w
Meaning	p	p ($\rightarrow$ q)	p, q	q

context. Following Enfield (2003), we can distinguish four stages in this process (table 10.1).

In the first stage, the original meaning of a word w is p. In the second stage, an inference q becomes more or less automatically activated together with p, perhaps as a result of strong correlations or of the context of use. In the third stage, the word means both p and q. At a possible fourth stage, the meaning p is forgotten.[15]

10.4.3 Syntactic Differences

The distinction between manner and result verbs is syntactically relevant: the two types of verbs differ in their argument patterns (Kaufmann, 1995; Rappaport Hovav & Levin, 2010, pp. 21–22).[16] To wit, the action described by a manner verb can be augmented, further specifying the event:

(10.7)   Oscar steamed the tablecloth clean/flat/stiff.

Here "clean/flat/stiff" describes the result of the action in different domains.

In contrast, result verbs cannot be augmented with a subevent from another domain (Rappaport Hovav & Levin, 2010; Croft, 2012a, p. 297):

(10.8)   *Kelly cleaned the dishes valuable.

(10.9)   *Tracy broke the dishes off the table.

(10.10)   *Oscar froze the people out of the room.

However, so long as the augmentation of the result stays within the domains that are strongly correlated with the result domain—and thereby expresses changes that are *expected*—it is acceptable:

(10.11)   Oscar froze the ice cream solid.

(10.12)   Tracy broke the vase into pieces.

(10.13)   Tracy broke the dishes against the table.

The reason that (10.13) works is because "against" is a force preposition that specifies the direction of the force pattern involved in "break."[17]

A similar argument also explains Kiparsky's (1997, p. 23) examples:

(10.14)   John pushed the cart, but it did not move.

(10.15)   *John rolled the cart, but it did not move.

The difference between (10.14) and (10.15) is that "push" is a manner verb that does not require any particular result (though there may be expectations). In contrast, "roll" is a result verb involving motion, so the movement cannot be excluded.

10.4.4   Semantic Implications

Result verbs describe the changes in the properties of the patient but do not entail how the changes are brought about. Levin and Rappaport Hovav (2010, p. 222) present the following example:

(10.16)   I cleaned the tub by wiping it with a sponge/by scrubbing it with steel wool/by pouring bleach on it/by saying a magic chant.

It shows how a result can be brought about in several manners besides the conventional one. Although result verbs generate conventional expectations about the corresponding manner, they do not entail them. Conversely, manner verbs do not entail results, although there are general expectations. *Wiping* normally leads to *wiping clean*, but the statement "I wiped the table but none of the fingerprints came off" (Rappaport Hovav & Levin, 2010, p. 22) is perfectly acceptable. This absence of entailments—from manner to result and from result to manner—is explained by the fact that the regions of forces patterns that represent the manner verbs are not strongly correlated with the regions for the result verbs.

The vectorial analysis from the event representations also explains why many result verbs have *antonyms* ("come-go," "cool-heat," "grow-shrink," "fill-empty," "dry-wet," "find-lose," "embark-disembark"). In particular, for any one-dimensional result domain, a verb referring to a vector representing a change in one direction can be complemented by a vector going in the other direction—provided the change process is reversible. (If it is not reversible, there can be no such verb: for example, there is no "uncook" or "unhit.") Of course, not all reverse vectors may be lexicalized. In contrast, very few manner verbs refer to force patterns that are reversible directed vectors, and consequently antonyms are rare among these verbs. One example, though, is "push-pull," where the verbs represent such a pair of one-dimensional force vectors (Zwarts, 2010b).

My analysis explains why many result verbs can be turned into adjectives: "the broken window," "the opened door," "the painted face."[18] In

these cases, the verb ("break," "open," "paint," etc.) is a result verb referring to a result vector that goes from outside a domain region to inside it
(from not broken to broken, from closed to open, from unpainted to
painted, etc.). The adjective simply expresses the region of the endpoint
of the result vector. In contrast, this mechanism does not work for manner
verbs, for example, "*the hit door," "*the pulled sledge," where no such
endpoint region can be identified.[19]

Rappaport Hovav and Levin (2010, p. 28) suggest that the semantic
difference between the two categories is that "all result roots specify
scalar changes, while all manner roots specify nonscalar changes."
They describe a scale as "a set of degrees—point of intervals indicating
measurement values—on a particular dimension." Thus they use the
notion of a dimension to characterize the difference between manner and
result verbs.

Their proposal has several problems, however, that are avoided by my
use of domains and the single-domain thesis. First of all, Rappaport Hovav
and Levin must allow two-point scales, which is not much scalarity (e.g.,
the scale for "arrive" is binary). Second, domains are more appropriate than
dimensions: for example, "paint" and "color" are result verbs that express
changes in the three-dimensional domain of colors. Third, I do not see
why manner verbs cannot be scalar, in particular when the force vector is
one-dimensional, as with "push."[20]

One class of result verbs is notably problematic for the scalarity hypothesis: it includes the verbs that describe change in the structure of an object,
for example, "break," "cut," "explode," "burn," "eat," and "melt." These
verbs do not represent "scalable" domains—unless binary scales are
allowed. Some of them, like "break" and "cut," express changes in the
topological properties of objects, such as connectedness, or changes in
meronomic structure. Other verbs in this class, such as "glue," "couple,"
and "dovetail," go in the other direction and connect parts into wholes.
All these verbs express higher-level change that is expressible in basic
domains. I will not elaborate here on how to extend the domain analysis
to include these cases.[21]

Rappaport Hovav and Levin (2010) consider only nonstative verbs in
their classification—presumably because their scalarity criterion does not
apply to stative verbs. In my analysis, stative verbs are a special case of
result verbs where the result vector is the identity vector, corresponding
to a point in some property domain.[22] Thus my theory handles these verbs
too. Even if the result vector is a point (the identity vector), it does
not follow that the force vector is zero, only that it is balanced by some

counterforce. There are stative verbs that express this kind of balance, for example, "stay" and "remain" (Talmy, 1988).

Semantically, there are strong similarities between stative verbs and adjectives. Adjectives in predicative use assign a static property to an object and in a comparative use make static comparisons between two objects. A result verb expresses a dynamic comparison of a property of an object with *itself*, before and after a change. Stative verbs behave like adjectives, since they describe a limiting case of change—from a region in space to itself, which is tantamount to just specifying a region (as is the case for adjectives). In many languages, the copula construction can be seen as transforming an adjective into a stative verb. In support of the idea that stative verbs are cognitively simpler than nonstative verbs, Gennari and Poeppel (2003) showed in their experiments that stative verbs demand less processing time.

10.5   The Role of Instruments

When discussing how the thematic roles of agent and patient are represented, I distinguished between the force vector of the event applied to the patient and the force vector generated by the agent. When an instrument is involved, the force exerted by the agent will be modified by the instrument and is thus different from the force vector affecting the patient.

The difference between the two force vectors shows up linguistically: the causal chain of John kicking the ball and the ball hitting the window can be expressed by

(10.17)   John hit the window with the ball

but not by

(10.18)   *John kicked the window with the ball.

Similarly, the causal chain of Mary lighting the fire and the fire heating the water can be construed as

(10.19)   Mary heated the water with the fire

but not as

(10.20)   *Mary lit the water with the fire.

The upshot is that whenever there is an instrument, the force vector applied to the patient (not the one applied by the agent) is the one that is primarily expressed. This accords with the proposed model of events.

The prototypical agent is volitional, while instruments are nonvolitional. Yet in English the instrument can be expressed as the subject:[23]

(10.21)   The hammer hit the nail.

In this sentence "hammer" is put in focus and functions as an agent. The analysis is supported by the inability to add a typical agent to the construction:

(10.22)   *The hammer hit the nail by Oscar.

10.6   Intentional Verbs

The prototypical action is volitional, and thus most manner verbs presume a volitional agent (Dowty, 1991; Croft, 2012a, p. 282). The typical meaning can then be metaphorically extended to a nonvolitional agent as, for example, when "touch" is extended from "Oscar touched the screen" to "The airplane touched the power line." In this example, agentivity is added to the domain matrix for "airplane."

A stronger assumption is that the agent is *intentional*; that is, the agent has a representation of a goal that it wants to obtain by acting. The distinction between volitional and intentional sometimes shows up in result verbs. In some cases, a special verb is used to mark an intentional result in contrast to another more neutral verb. The classical case is "kill" versus "murder." "Murder" is intentional, while "kill" is undetermined with respect to intentionality. Thus "murder" cannot occur with nonintentional agents (Levin & Rappaport Hovav, 2005, p. 27):

(10.23)   *The explosion murdered Larry's neighbor.

Another example is the distinction between "blink" and "wink." A blink is an (often unintentional) action, that is, a pattern of forces exerted on the muscles around the eye. In contrast, a wink is an intentional action, combining the action of blinking with the goal "to awaken the attention of or convey private intimation to [a] person" (*The Oxford Concise Dictionary*).

Many events involving goals can be construed from either of two perspectives: the physical action on an object or the intentional action leading to the fulfillment of a goal.[24] Such a situation can still be expressed with the aid of a single verb, since the fulfillment of the intention *presupposes* a physical action. Important examples include "give," "buy," and "sell." All involve (at least) three entities: agent, object, and recipient. The intentional aspect of such events concerns object ownership (or, more generally, being in control of the object) and the physical properties of those objects (typically a movement of the object).

In most cases, the difference between the intentional and the nonintentional use is not marked by a special verb:

(10.24)   Oscar baked the potatoes for an hour.

(10.25)   Oscar baked a cake.

Atkins, Kegel, and Levin (1988) distinguish between two senses of "bake": (i) to change the state of something by dry heat in an oven, exemplified by (10.24); and (ii) to create by means of changing the state of something by dry heat in an oven, exemplified by (10.25). Example (10.25) is a meronomic change, as discussed in section 6.3. In sense (i), "bake" is a manner verb. In sense (ii), it seems to cover both manner and result—seemingly contradicting the single-domain thesis. The contradiction, however, is only apparent. Note that sense (ii) expresses an intentional event involving the creation of an object. Once again, the intentional construal presupposes the physical, so the two construals can be summarized by the intentional. The meaning (ii) is a result verb that can be seen as derived from the manner verb of meaning (i). The same ambiguity of meaning can be found in verbs such as "cut," "brush," "chop," "grind," and "mow." Consequently I propose two meanings for these verbs, involving one intentional and one nonintentional reading.[25]

The two meanings behave differently in English: (a) the potatoes are baked (accusative use); (b) the potatoes bake (unaccusative use); (c) the potatoes are baked by John (accusative use); (b) *the potatoes bake by John (unaccusative use).[26]

Similarly, "weigh" has one intentional reading (determining the weight of something) and one nonintentional (having a certain weight). This explains the difference between "The grain is weighed by Tim" (intentional) and "*250 pounds is weighed by Tim" (nonintentional) (Croft, 1991, p. 8).

Goldberg (2010) argues that *accomplishments*—which she defines as "predicates that designate both an activity and the end-state of that activity"—involve two causally related subevents (see also Dowty, 1979; Rappaport Hovav & Levin, 1998).[27] One example is "fill": "to infuse until full." In my analysis, these verbs are intentional and can therefore generate construals of two subevents: the intended result and the manner of achieving the result.[28] Since the intention presumes an action, the verb summarizes the two subevents. "Fill" can also be used for an activity, as in "She filled the tank for ten minutes." In this case, the verb only denotes the manner of acting, without implying any intentional result.

The upshot is that verbs involving intentional actions are not really counterexamples to the single-domain thesis. However, the strong coupling between the two construals creates the illusion that these verbs describe both manner and result.

The construal involving the physical action is primary to the intentional construal. The primacy of manner verbs shows up when the (telic) intentional construal is complemented with a (nontelic) modifier that applies to the physical action:

(10.26) *Oscar baked Victoria a cake for an hour.

Kiparsky (1997) argues that denominal verbs refer to generically intentional activities. He gives the following examples:

(10.27) *The explosion painted the workers red.

(10.28) *Velázquez painted the brush red.

The primary meaning of "paint" is to cover with paint intentionally. Since an explosion cannot have an intention, (10.27) is thus not acceptable. Similarly, when Velázquez in (10.28) dips his brush in the paint, it is covered with red paint, but that event is not part of his intention. There exist more or less metaphorical uses of "paint" that involve inanimate agents, though:

(10.29) The sun painted the sky pink, orange, and purple.

Kiparsky (1997) presents lists of locatum verbs (e.g., "fuel," "pepper," "saddle") and location verbs (e.g., "bottle," "pen," "shelve") that all involve intentions. He argues that when an object is used to generate a verb, it involves a canonical use of the object (also cf. McIntyre, 2007, pp. 7–8). Thus:

(10.30) *The motels were full, but the authorities managed to imprison all the victims of the flood.

Using a prison to house victims is not a canonical use of the prison, and thus (10.30) is odd.[29]

Kiparsky (1997, pp. 12–16) notes a contrast between, on the one hand, "true" denominal verbs such as "box" where things must be put in the corresponding noun, such as

(10.31) *She boxed a present in a brown paper bag.

and, on the other hand, "apparent" denominal verbs such as "shelve" where "the nominal meaning is to varying degree attenuated or 'bleached'" (p. 13):

(10.32)   He shelved a book on the windowsill.

Kiparsky's analysis of this (p. 14) is that the noun "shelf" means "thin flat narrow horizontal elevated surface," and the denominalized verb "shelve" means "to put on a shelf-like thing." My analysis of the apparent denominalized verbs is instead that they involve the *force patterns* related to the things expressed by the noun. Hence the verb "shelve" means "to put an object on a thin flat narrow horizontal elevated surface that counterbalances the force of the object." Similarly, verbs like "hammer" and "brush" refer to the force patterns used with hammers and brushes rather than the objects themselves (examples from Kiparsky, 1997, pp. 15–16):

(10.33)   He hammered the desk with his shoe.

(10.34)   He brushed his coat with his hand.

From a developmental point of view, it can be argued that it is easier for a child to understand that it acts to obtain a goal than to understand the goals and intentions of others (Huttenlocher, Smiley, & Charney, 1983). In contrast, the actions of others are easily observable and can be categorized at an early age. This difference shows up in the way that children produce and understand verbs. Huttenlocher et al. (1983) show that when it comes to the comprehension of manner verbs, describing actions are understood at an earlier age than result verbs.[30] On the other hand, regarding production, children around two years produce result verbs such as "open," "come," and "break" much more frequently than manner verbs such as "turn," "ride," and "play" in relation to their own actions. However, in relation to the actions of others, result verbs are less frequent and mainly used in requests ("open door"). This is supported by the fact that parallels in actions between the subject and others are understood at an earlier state of development than parallels involving goals and intentions (see sec. 3.5).

10.7   Perception Verbs

I next turn to verbs related to perception.[31] For these verbs, the "forces" involved will be of a different nature than in typical cases. In the perception event expressed by "I hear the owl," "hear" is a result verb that describes a state of the subject, who functions as the patient in the event (Jackendoff, 2007, p. 205). In this case, the subject does not exert a force, that is, use its effectors: instead the subject is changed through its detectors.[32] In such a situation, the relevant force vector is the action of the sound on the agent.

In contrast, "I *listen* to the owl" implies an active directing of attention toward the owl. The subject is thus an agent in the prototypical sense.[33] Directing one's attention is construed as a case of exerting a force. Accordingly, "listen" is a manner verb.[34] A similar distinction can be made between "see" (a result perception verb) and "look" (a manner verb) and between "feel" (a result perception verb) and "touch" (a manner verb). "Smell" seems to be ambiguous, alternatively expressing the result and the manner meaning.[35]

10.8   Concluding Remarks

Within linguistics, much work in semantics derives from generalizations of some limited set of linguistic data, which are then used to open windows onto underlying cognitive phenomena. Throughout the book, I follow the reverse path, starting from a general cognitive framework to derive its implications for semantics. I have extended the theory of conceptual spaces to models of actions and events. The models have been applied to suggest significant cognitive and communicative constraints on lexicalization processes.

The main semantic thesis in this chapter is that verbs refer to convex regions of vectors defined by a single semantic domain (in parallel to adjectives that refer to convex regions of a single domain). Together with the framework of conceptual spaces, this approach has allowed me to explain many features of the semantics of verbs. Here I will just summarize some of the main points. First, the models of actions and of property change make it possible to predict both the similarity of meanings between verbs and the super/subordinate semantic hierarchy of verbs. Traditional semantic theories cannot achieve this in a natural way. Second, the manner/result distinction falls out immediately from the single-domain thesis. Third, I have highlighted the role of intentionality in verb meaning. I have argued that many verbs that seem to violate the single-domain thesis actually have dual lexical potential: one use that includes the intention of the agent, and one that includes only the manner of the action. Fourth, as was shown in the previous chapter, the vectorial approach provides a simple and natural model of the verb aspects proposed by Vendler (1957).

To sum up, the analysis of verbs presented in this chapter parallels the previous analyses of pointing and of adjectives and nouns. The analysis has mainly covered verbs that are related to actions and perceptual domains. As a consequence, I have not treated several classes of verbs, in particular auxiliary verbs and modal verbs. Modal verbs are similar to the coercion verbs discussed in section 10.3, since they always combine with another verb, and they are therefore of a different nature than the basic verbs.

The analysis of events brings out the semantic affinities between result verbs and adjectives. According to the theory of this book, their meanings can be represented by closely connected geometric structures. Adjectives (often together with a copula) can be seen as represented by a result vector of an event: "The soup *is hot*" is represented by a stative (zero) result vector and "The soup *becomes hot*" is represented by a result vector going from the cool to the hot region of the temperature domain. The verbs "is" and "become" function as general placeholders for stative and nonstative result verbs expressing changes of properties. Analogously, the verb "go," in combination with a prepositional phrase ("Oscar went to Berlin"), is often used as a placeholder for result verbs expressing changes of position. Similarly, "do" and "make" are placeholders for manner verbs. These general verbs can be seen as the top superordinate verbs in the verb classes generated from the event model. In parallel with pronouns, they function as "pro-verbs."

11 The Geometry of Prepositional Meaning

In most languages, prepositions form a closed class with a limited number of representatives. However, they are often used for a wide range of meanings.[1] This semantic flexibility makes it difficult to provide an exhaustive analysis of their semantics. The linguistic literature on prepositions is extremely rich, and I have no ambition to do it justice. In this chapter, I focus on the convexity of their meanings and on which domains are expressed by prepositions. I argue that for many prepositions, the force domain is central.[2] In contrast to many other analyses, I also defend the position that prepositions have a central meaning and that other meanings can be derived via a limited class of semantic transformations. Although I focus on prepositions, this principle extends to other word classes.

Before I turn to the semantics of prepositions, I make some general remarks about polysemy. The transformations I discuss apply to several of the topics of Part II of the book, not only to prepositions.

11.1 Semantic Transformations

11.1.1 Minimal versus Full Specification of Meaning

Not to mean one thing alone is to mean nothing.
—Aristotle

Many words seem to be polysemous; that is, they appear to have a number of different meanings that are only distantly related. There are two basic ways of handling this phenomenon in a semantic theory (Lakoff, 1987, p. 420; Tyler & Evans, 2001, pp. 727–733; Zlatev, 2003; Van der Gucht, Klaas, & De Cuypere, 2007). One is *full specification*, where each meaning of a word is represented separately in the lexicon, but semantic relations between the different meanings can also be specified. The other is *minimal specification*, where one meaning of a word is considered to be central, and other meanings are derived from the central one by additional information from the context or by semantic transformations.[3]

Lakoff (1987, p. 422) argues in favor of the full specification interpretation, and Tyler and Evans (2001, pp. 731–737) argue for a weaker form they call *principled polysemy*, the main difference being that Tyler and Evans provide criteria for identifying separate meanings.[4] Principles of cognitive economy, however, provide strong arguments against the full specification position. Jackendoff (1983, pp. 118–189) writes: "The mind does not manufacture abstract concepts out of thin air. . . . It adapts machinery that is already available, both in the development of the individual organism and in the evolutionary development of the species." Our semantic memory would be strained if we were to have separate encodings of all the twenty-four meanings of "over" that Lakoff (1987) identifies and of corresponding meaning networks for other words. In contrast, remembering a prototypical meaning and then using some general semantic principles for creating other meanings would be more economical for our memory.

For this reason, I endorse the minimal specification approach. The burden on me will be to specify what ways exist to create new meanings from a central element. My primary tool for this is what I call semantic transformations.

11.1.2 Main Transformations

I will use the preposition "over" to illustrate some of the main transformations of semantic content that are needed to defend a minimal specification position. My position is that there are only a few basic transformational mechanisms. In section 2.6, I identified two main categories of transformations: *metonymy* as a (re)focusing of attention (within a domain) and *metaphorical mapping* (between domains). Note that I use both metonymy and metaphor in a broader way than is usual within linguistic theory.

At the end of his paper, Dewell (1994, pp. 375–376) presents a list of the semantic transformations that he finds necessary for the analysis of "over." Many of them—*segment profiling, trajector-part profiling, extending-path*, and their subcategories—can be seen as special cases of *attention focusing* of the kind that I discussed in section 1.3, and thus they fall under my broad characterization of metonymy. The *multiplex-mass* transformation was also presented there as a case of zooming out. As a variation of this, Dewell also considers *multiplex path*, where a pointlike trajector moves in a variety of directions that can be construed as a mass. His *resulting state* is what is here called endpoint focus. I have analyzed endpoint focus, which again was categorized as a shift of focus in section 1.3, as a special case of metonymy.

In addition to the transformations generated by the attentional changes, as described in section 1.3, and other metonymical operations, I discuss several examples of *metaphorical* transformations in the example of how "over" is used, for example, by mapping from the force to the time domain. What Dewell (1994, p. 361) calls *shifting perspectives* are the dimensional changes from vertical to horizontal (or reversing the directions of the vertical) that I classify as a form of metaphorical transformation. A full analysis of the different submechanisms is, however, beyond the scope of this book. Such an analysis would be useful not only for lexical research but also for communication modeling.

11.1.3 Iterated Transformations

Semantic transformations operate on several levels in language use: they are partly lexicalized, partly introduced by coordination in communication, and partly by context. As an example of the role of context, consider Lakoff's (1987, pp. 429–439) analysis of "rotated" schemas for "over."

(11.3)   Superman flew all over downtown Metropolis.

(11.4)   Harry climbed all over the canyon walls.

(11.5)   *Superman flew all over the canyon wall.

Lakoff considers (11.3) and (11.4) to be acceptable sentences, but not (11.5). He thinks that the rotational transformation involved in using "over" with a vertical landmark requires contact between the trajector and the landmark. However, this conclusion depends on the salience of direction of the landmark in the context. If Harry has been climbing all over the canyon wall the whole day looking for an edelweiss flower, the verticality of the world has become extremely salient. He may then sigh and say, "I wish I were Superman. Then I could fly all over the canyon wall." Once the rotational transformation has been established, it becomes easier to add another transformation and still be understood.

Here I will focus on the lexicalization of transformations rather than on their dependence on the context. Let me begin with the example "topless district" (Fillmore, 1978). Here it is not the district that is topless, or the bars in the district, or the waitresses who work in the bars, but the dresses that the waitresses wear. What has happened in this lexicalization process is a series of three *totum pro parte* transformations. First, the waitresses are taken to stand for the dresses they wear; second, the bars stand for the waitresses who work in them; and third, the district stands for the bars in it. The lexicalization must occur in a stepwise fashion: one cannot go directly from "topless dress" to "topless district" without the intermediary metonymies.

One can also find iterations of metaphors and combinations of metaphors and metonymies. Consider the French "avoir la tête en feu" (have one's head on fire). Here the head is a *pars pro toto* metonymy for a person combined with the fire as a metaphor for the person being excited or agitated (see De Lucs, 1993).

In general, applying transformations to a prototypical meaning of a word will generate a graph structure of meanings that are connected to each other via various transformations. As a matter of fact, I believe that the twenty-four meanings that Lakoff (1987) and the fourteen meanings that Tyler and Evans (2001) present can all be analyzed as combinations of shifts in attentional foci, metaphorical transformations, and elaborations (superimpositions). Furthermore, the graph structure generated by the combinations can also be interpreted as a semantic map that can be used to investigate the differences between how different languages partition meaning structures. As is well known, although two languages may have the same central meaning of a word, the set of transformations that are lexicalized may vary considerably, and thus the extended meanings of the word will be different. As Langacker (1991b, p. 3) writes: "There is no way to predict precisely which array of extensions and elaborations—out of all those that are conceivable and linguistically plausible—have in fact achieved conventional status."

Meaning transformations come at different *cognitive costs*. Some metaphorical mappings, for example, may be more difficult to understand than other more conventional ones.[5] Psycholinguistics uses a number of methods—for example, measuring reaction times and error frequencies—to compare the cognitive efforts of comprehension or production tasks.

Considerations of cognitive costs motivate the general constraint on iterated semantic transformations that the set of lexicalized meanings of an expression should be *connected*. This means that if word w originally has a meaning m_1 and this meaning has been extended by a series of transformations to meaning m_n through meanings m_2, m_3, . . . , and meaning m_n also falls under the word w, then all meanings m_2, m_3, . . . , also fall under w.[6]

On the basis of these general remarks on semantic transformations, I now turn to how they can be applied to account for the semantics of prepositions.

11.2   The Semantics of Prepositions

First of all, we must consider the communicative function of prepositions. Let me return to the paradigmatic example of identifying a referent. Adjectives modify a noun by specifying a property in a domain belonging to the object category that is referred to. Most prepositions can be grouped into two classes: *locative*, indicating where something *is*, and *directional*, indicating where something *is going*.[7] Locative prepositions modify a noun (noun phrase) by specifying the *location* (a region) in the visuospatial domain: "Give me the bottle *behind* the bread!" This function is required for communication on level 1—instruction—and is similar to the function of adjectives. Another function is fulfilled by directional prepositions. In a sentence such as "Oscar went *to* the library," the phrase "to the library" has the same function as a result verb: it specifies the *result vector* of an event. This function is mainly relevant on level 2—coordination of inner worlds. Note that in both functions, the preposition is combined with a noun (or a noun phrase).

In line with the analyses of the semantics of adjectives and verbs, I tentatively put forward the following thesis:

Single-domain thesis for prepositions: Prepositions represent convex sets of points or paths in a single domain.

As I will show, locative prepositions are represented by sets of points and directional prepositions by sets of paths.

In parallel to the single-domain thesis for verbs, the claim is that each use of a preposition builds on a single domain, but it is not required that all uses are based on the same domain. For example, I will argue that most typical uses of the prepositions "over," "on," and "in" depend on the force domain. However, there are common metaphorical transformations of meanings that bring these prepositions into the visuospatial domain. As a matter of fact, metaphorical uses of prepositions are ubiquitous. Nevertheless, I will argue that for each preposition there is a central meaning that depends on a primary domain.

11.3   Spatial Representation Using Polar Coordinates

11.3.1   Polar Coordinates and Convexity

To model the meaning of prepositions, I need to make some assumptions about how to model the visuospatial domain. Normally this domain is represented with the aid of the Cartesian coordinates x, y, and z, representing width, depth, and height, and where distances are measured using a Euclidean metric. However, another way of representing space may be cognitively more realistic, namely, in terms of *polar* coordinates, which represent points in space in terms of distance and angles. Cognitively, a polar representation of space is more natural than a Cartesian one, since our visuospatial perceptual system is made for estimating directions and distances from ourselves rather than estimating distances between two points outside us.[8] We are so influenced in our culture by Euclidean geometry, Cartesian coordinate systems, Newtonian mechanics, and Kantian a prioris that we have difficulties seeing that there are other ways of describing spatial perception.

I start with a three-dimensional space S defined in terms of polar coordinates. It is assumed that the space has an *origo* point *o*. A point *p* is represented as a triple $<r, \varphi, \theta>$ where:

r (the *radius*) is a real number (with $r \geq 0$) representing the distance of *p* from the origo.
φ (the *azimuth angle*) is the angle (with $0° \leq \varphi < 360°$) between *p* and the "north" axis, perpendicular to the zenith (the *azimuth*)
θ (the *polar angle*) is the angle (with $0° \leq \theta \leq 180°$) between *p* and the "upward" axis (the *zenith*)

Following common practice in the use of polar coordinates, two absolute frames of reference are already built into the polar coordinates, namely, the zenith (up) and the azimuth (north), as the fixed reference directions relative to which other angles are defined. (In most cases, the upward direction is determined by gravitation.) Notice that when the polar angle is 0° or 180°, then the value of φ is arbitrary. I assume that the angle φ goes clockwise when seen from above, so that east is 90° and west is 270°.

Given the representation of polar coordinates, I can define a notion of *polar betweenness* that is different from the one generated by the standard Euclidean metric:

A point $b = <x_b, \varphi_b, \theta_b>$ lies between a point $a = <x_a, \varphi_a, \theta_a>$ and a point $c = <x_c, \varphi_c, \theta_c>$ if there is some k, $0 < k < 1$ such that $x_b = kx_a + (1 - k)x_c$, $\varphi_b = k\varphi_a + (1 - k)\varphi_c$ iff $|\varphi_a - \varphi_c| \leq 180°$, and $\varphi_b = k\varphi_a + (1 - k)(\varphi_c - 360°)$ iff $|\varphi_a - \varphi_c| > 180°$, and $\theta_b = k\theta_a + (1 - k)\theta_c$.

The azimuth angle also takes on values greater than 180°, but betweenness is defined with respect to the smallest angle.

The polar coordinates introduce a different metric on the space, compared with the standard Euclidean metric. Consequently the "lines"

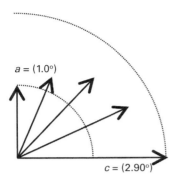

Figure 11.1
Polar betweenness.

generated by this polar betweenness relation will be "curved," if seen with Euclidean glasses.[9] This is illustrated in figure 11.1 for three vectors between the vectors $a = <1,0°>$ and $c = <2,90°>$. These vectors are between, because their radiuses are between those of a and c, and at the same time their angles are between those of a and c.

A region R in S is then defined to be *polarly convex* if and only if for all points a and b in R, any point c that is polarly between a and b is also in R.

One thing to note about this definition, in comparison to Euclidean convexity, is under what coordinate transformations convexity is preserved. Euclidean convexity is preserved under multiplications, translations, and rotations of the coordinate system. In contrast, polar convexity only preserves convexity under multiplications (changing the values of the r-axis) and rotations (changing the values of the φ- and θ-axes). If a translation occurs, that is, if the origo moves, then convexity may not be preserved. It should also be noted that the only polarly convex set that allows both unrestricted multiplication and unrestricted rotation is the full space. The relevance of this will be apparent later.

In this context, Talmy (2000, chap. 1) proposes a topology principle that "applies to the meanings—or schemas—of closed-class forms referring to space, time, or certain other domains." This principle, which he claims is a language universal, says that distance, size, shape, or angle from such schemas play no role for the meaning of these forms. He illustrates with the preposition "across," which "prototypically represents motion along a path from one edge of a bounded plane to its opposite. But this schema is abstracted away from magnitude. Hence, the preposition can be used equally well in 'The ant crawled across my palm,' and in 'The bus drove across the country.' Apparently, no language has two different closed-class forms whose meanings differ only with respect to magnitude for this or any other spatial schema." This argument means that prepositions have configurational meanings in the sense of section 2.8.2.

11.3.2 Motion along Paths

The previous section introduced the formalism for representing the location of a trajector in terms of one vector. If a trajector is moving or if it is extended in shape, then the notion of a *path* is needed (see, e.g., Jackendoff, 1983; Talmy, 2000; Eschenbach, Tschander, Habel, & Kulik, 2000; Zwarts, 2005; and many others). There are different ways to represent a path, but I adopt the more common way of representing it as a directed curve, that is, as a continuous function p from the real interval [0,1] to S. The values of the interval [0,1] do not represent moments of time, but they are an ordering mechanism. What is important is that the path represents "locations in sequence," so to say.

The starting point of a path p can be denoted as $p(0)$, the endpoint as $p(1)$, and for any $i \in (0,1)$, $p(i)$ is an intermediate point. All of these will be points that are represented in terms of polar coordinates. It will be convenient later to refer to these coordinates in the following ways:

radius$(p(i))$ is the radius of the path p at i.
polar$(p(i))$ is the polar angle of p at i.
azimuth$(p(i))$ is the azimuth angle of p at i.

I also assume that the path is *simple*, that is, it does not cross itself. This can be defined by saying that for all i and j, $p(i) \neq p(j)$.

11.4 Locative Prepositions

A preposition describes a relation between a trajectory and a landmark ("between" is an exception, since it involves a relation between a trajectory and two landmarks). The landmark will usually not be point sized, but it will occupy an extended *region* of space, its *eigenplace* (Wunderlich, 1991). I assume a function *loc* that assigns a subset $loc(x) \subseteq S$ as an eigenplace to every convex object x. For convenience, I often simply designate this eigenplace as x and refer to it as "landmark," though it is strictly speaking $loc(x)$ and "eigenplace of landmark."

When one judges the relation between a trajectory and a landmark, the center of the landmark will function as the origo of S. To make the mathematics not too complicated, I consider only *convex* eigenplaces and restrict the analysis to circular landmarks with the origo in the center. It is an idealization that all landmarks have this shape, but for the meanings of the locative prepositions, this idealization does not result in any major deviations.

11.4.1   Regions for Locative Prepositions

I now show that by using polar coordinates, locative prepositions can be given a highly systematic description that brings out more explicitly the spatial features of each preposition. A basic distinction in the system of prepositions is between internal and external regions, corresponding to the prepositions "inside" and "outside," respectively. These regions can be defined as sets of points, where r is the radius of the landmark. For simplicity's sake, these regions are restricted to the horizontal plane by including only the horizontal angle in the coordinates.

inside: $\{<x, \varphi>: x < r\}$
outside: $\{<x, \varphi>: x > r\}$

The corresponding regions can be diagrammed by shading the area where the endpoints of the vectors are (fig. 11.2).

The three coordinates of a polar system provide three ways of dividing the space *outside* a landmark. The first is *distance* from a landmark (the x-coordinate). A first distinction is to divide the space into points that are *near* or *far* (fig. 11.3). Of course, at what distance the division is made depends on the context, in particular the size of the landmark: what is *near*

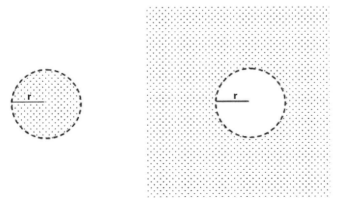

Figure 11.2
Inside and *outside* regions.

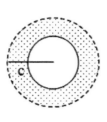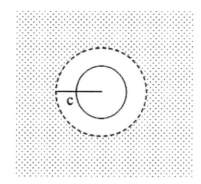

Figure 11.3
Near and *far* regions.

the sun in space covers a much larger distance than what is *near* a golf ball on the ground. Moreover, the division is not sharp but might allow for a vague gap between *near* and *far*.

The denotations of *near* and *far* can be given in the following way as sets of located polar coordinates, for a given landmark region with radius r and a contextually given norm c for distance.

near: $\{<x, \varphi>: r < x < c\}$
far: $\{<x, \varphi>: c < x\}$

Working in the horizontal plane are prepositions like "in front of," "behind," "beside," "to the left of," and "to the right of," and the cardinal directions "north of," "west of," "south of," and "east of." Much has been written about the different frames of reference that are used here, with different terminologies, such as egocentric and allocentric, relative and absolute, object-centered and viewer-centered (Levinson, 1996; Bohnemeyer, 2012). I follow the terminology of Levinson. The cardinal prepositions use an *absolute* frame of reference, directly tied to the fixed (north) reference direction of φ. The other prepositions use either an intrinsic frame of reference (based on features of the landmark itself) or a relative frame of reference (based on the position of an observer). I restrict the analysis to the intrinsic frame of reference by assuming that some angle $f \in \varphi$ is assigned to a landmark x that represents the front direction of $loc(x)$. This results in a number of "focal" directions from x in the horizontal plane, namely, $f, b = f + 180°, r = f + 90°, l = f - 90°$. The regions for the prepositions can then be defined in terms of closeness to these focal directions, leading to borders that are vague and partially dependent on context:

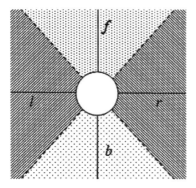

Figure 11.4
In front of, behind, to the left of, and *to the right of* as regions of the angle φ around a landmark.

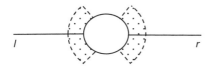

Figure 11.5
The region for *beside.*

in front of: {<x, φ>: r < x and φ is close to f}
behind: {<x, φ>: r < x and φ is close to b}
to the left of: {<x, φ>: r < x and φ is close to l}
to the right of: {<x, φ>: r < x and φ is close to r}

If the prepositions partition the horizontal angle into four regions, then this would look as in figure 11.4.

The preposition "beside" could be seen as the union of "left of" and "right of," but it seems to have an extra element of proximity, which is lacking in the other horizontal prepositions (Svenonius, 2012). For the time being, I assume that "beside" covers angles of φ that are close to $f \pm 90°$, but I take a closer look at "beside" in the next subsection.

I next turn to the preposition "between," which is, in its most prominent use, based on two landmarks (Habel, 1989; Van der Zee & Watson, 2004). In this case, the reference angle is determined by the line between the two landmarks. Each landmark can be seen as generating a cone going in the direction of the other landmark. The region representing "between" can then be defined as the set of points that belong to both cones.

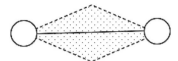

Figure 11.6
The region for *between*.

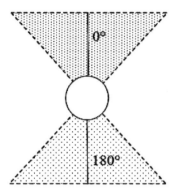

Figure 11.7
Above and *below* as regions.

The third division concerns the altitude coordinate θ that can be used to determine the regions for the prepositions "above" and "below." (I compare these to "over" and "under" in section 11.6.) The "prototypical" directions are 0° (straight up) and 180° (straight down). "Above" corresponds to values of θ that are close to 0°, and "below" corresponds to values that are close to 180° (see fig. 11.7, which should be seen as a two-dimensional cross section of the regions). Again, how close the values should be depends on the context and is a matter of vagueness. The prototypical meanings of the prepositions lie on the vertical axis.

The following definitions give us such regions for a landmark with radius r, given the appropriate notion of closeness:

above: $\{<x, \varphi, \theta>: x > r, \theta$ is close to 0°$\}$
below: $\{<x, \varphi, \theta>: x > r, \theta$ is close to 180°$\}$

In summary, dividing the space into regions along the three polar coordinates generates the regions for most of the common locative prepositions in a natural way, by imposing simple conditions on the magnitude of coordinates.

11.4.2 Convexity of Locative Prepositions

Given the notion of polar betweenness, the question is now whether regions of locative prepositions are polarly convex. The answer is clearly affirmative for the regions of the "angular" prepositions in the horizontal plane, namely, the ones that have one single "cone" or halfspace: "in front of," "behind," "to the left of," "to the right of," "north of," "above," "below," and so on. They are all polarly convex according to the definition in (2). Since intersections of convex regions are also convex, "between" satisfies the convexity principle. The reasoning is as follows. If the "cones" of the two landmarks are already convex, then the region that we create by intersecting these two regions (given the appropriate notion of intersection for regions of polar coordinates) will not create any discontinuities that violate convexity. It can be noted that the regions for all these prepositions are all closed under multiplication along the r-axis (magnitude), but not under rotation.

Next consider "outside," "near," and "far," which prima facie seem problematic for convexity, because there is a gap in the center of the region, where the landmark is, for all three prepositions, and, for "far," the area that is near the landmark. If position a to the east of my house is outside (near, far) and a position b to the west is outside (near, far), then there are definitely positions in between that may not be outside (near, far). But this description depends on Euclidean betweenness. If polar betweenness is applied to these regions, then they are convex. The curved nature of the line between two points a and b leads around the gap in the middle. Hence a point in the middle of a to the east and b to the west is a point either to the north or to the south. This is illustrated in figure 11.8.

Similarly, "inside" is straightforward, given the idealizing assumption about landmarks I have made. Note also that, in contrast to the previous class of prepositions, the regions for "outside," "inside," "near," and "far" are all closed under rotations, but not under multiplication.

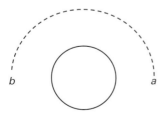

Figure 11.8
Polar betweenness "around" a gap.

I am left then with the convexity of the preposition "beside." A natural description is that the value of the angle φ of the vector going from the landmark does not deviate considerably from either the left (−90°) or the right (+90°) direction of y. In figure 11.5, it seems that this set of vectors consists of two separate sets, which would then violate the convexity requirement. One conclusion might be that "beside" is in fact a nonconvex preposition, maybe because it has a *disjunctive* definition: vectors that are close to +90° *or* to −90°. While intersections of convex regions are convex, unions are not necessarily so. However, another conclusion might be that I have misanalyzed "beside."

Are there ways to analyze it as convex? One possibility would be to say that "beside" covers all the horizontal directions, but its use for the forward and backward direction is preempted or blocked by the prepositions "in front of" and "behind." Another possibility would be to restrict "beside" to just one side at a time, as if we are saying "at *a* side."[10]

In fact, there are not many complex prepositional phrases that violate convexity, only some artificial cases like "diagonally above" and "exactly one or exactly two meters above." Even run-of-the-mill modified cases like "two feet above" or "far outside" are convex (given my assumptions about landmarks). Not only basic prepositions are convex, but even many of the complex ones. This means that convexity gains additional support as a general semantic constraint for locational prepositions. Of course, the principle should also be analyzed cross-linguistically, but that is beyond the scope of this chapter.

11.5 Directional Prepositions

I now consider prepositions that are used to express how a trajector moves relative to a landmark, which are the subject of much study (a recent example is Pantcheva, 2010). They include the following:

goal prepositions: *to, into, onto, toward*
source prepositions: *from, out of, off, away from*
route prepositions: *through, over, along, around, across*

In addition, it is possible to get directional readings for locative prepositions:

goal: (to go) *under, behind,* . . .
source: (to come) *from under, from behind,* . . .
route: (to pass) *under, behind,* . . .

The source needs to be marked by the source preposition "from," while the possibility to get a goal or route interpretation depends very much on the verb and other factors (see, e.g., Gehrke, 2008; Nikitina, 2008).

11.5.1   Representing Directional Prepositions as Sets of Paths

Most of the directional prepositions can be represented as imposing a locative condition on a particular part of the path, for instance, on the endpoint (goal) or starting point (source). Stricter and weaker definitions are possible, as discussed in Zwarts (2005), but here I use definitions that only involve opposite conditions on the starting point and endpoint:[11]

Goal prepositions

to = {p: near($p(1)$) and not near($p(0)$))}
into = {p: inside($p(1)$) and not inside($p(1)$))}
(to) behind = {p: behind($p(1)$) and not behind($p(0)$))}

Source prepositions

from = {p: near($p(0)$) and not near($p(1)$))}
out of = {p: inside($p(0)$) and not inside($p(1)$))}
from behind = {p: behind($p(0)$) and not behind($p(1)$))}

Route prepositions

by, past = {p: not near($p(0)$), not near($p(1)$), and there is an $i \in (0,1)$ such that near($p(i)$)}
through = {p: not inside($p(0)$), not inside($p(1)$), and there is an $i \in (0,1)$ such that inside($p(i)$)}
(via) behind = {p: not behind($p(0)$), not behind($p(1)$), and there is an $i \in (0,1)$ such that behind($p(i)$)}

Other prepositions compare the endpoint with the starting point:

toward = {p: *radius*($p(1)$) < *radius*($p(0)$) and for all $i \in [0,1]$ *radius*($p(i)$) > r}
away from = {p: *radius*($p(0)$) < *radius*($p(1)$) and for all $i \in [0,1]$ *radius*($p(i)$) > r}

I next come to a class of directional prepositions that are at a level of greater complexity than the others, namely, "around," "across," and "along." Notice that these are also morphologically complex, derived from expressions of shape or orientation like "round," "cross," "long." What is different from the other route prepositions is that they are specified not in terms of a location at a particular point of the path but in terms of the shape or orientation of the path as a whole. *Around* paths are round in some sense, *across* paths are orthogonal to the main axis of the landmark, *along* paths parallel.

I have to set aside "across" and "along," because they involve landmarks that are elongated, which goes beyond my simple model of the landmark as a region around an origin. This leaves "around," which is a preposition with a wide range of meanings (Zwarts, 2003). A very strict interpretation of "around" (restricted to the horizontal plane) takes it as corresponding to a full and perfect circular path:

around = $\{p$: there is an $r > 0$ and an $\alpha \in (0°, 360°)$, such that $p(i) = (r, \alpha \pm i\varphi)\}$

11.5.2 Convexity of Directional Prepositions

It is not obvious how to define betweenness for path. If I focus on simple paths that are defined as mappings from the interval [0,1] to points $<x, \varphi>$ in the horizontal surface, there is a solution that follows the standard definition: Let $p_1(i)$ and $p_2(i)$, where $0 \le i \le 1$, be two functions mapping onto $<x_1(i), \varphi_1(i)>$ and $<x_2(i), \varphi_2(i)>$ respectively. Then the path $p_3(i)$ is said to be *polarly between* $p_1(i)$ and $p_2(i)$, if and only if there is some k, $0 \le k \le 1$, such that for all i, $0 \le i \le 1$, $p_3(i) = <x_3(i), \varphi_3(i)> = k<x_1(i), \varphi_1(i)> + (1 - k) <x_2(i), \varphi_2(i)>$.[12]

Given these definitions, it is easy to prove that the meanings of the goal prepositions are convex. Convexity for "into" and "out of" follows from the convexity of the points in the landmark, which I have assumed to be convex. Similarly, the convexity of "to" follows from the fact that the region for "near" is convex. To be precise, if $p_1(i)$ and $p_2(i)$ are two paths such that near($p_1(1)$) and near($p_2(1)$), then for any path p_3 between p_1 and p_2, it will also hold that near($p_3(1)$). An analogous argument holds for "from," and similar arguments can be made for "(to) behind" and "from behind," since the region for "behind" is convex.

The convexity of "toward" is easy to prove: if $p_1(i)$ and $p_2(i)$ are two paths such that $radius(p_1(1)) < radius(p_1(0))$ and for all $i \in [0,1]$ $radius(p_1(i)) > r$, and $radius(p_2(1)) < radius(p_2(0))$ and for all $i \in [0,1]$ $radius(p_2(i)) > r$, then for any path p_3 between p_1 and p_2, it will also hold that $radius(p_3(1)) < radius(p_3(0))$ and for all $i \in [0,1]$ $radius(p_3(i)) > r$. An analogous argument shows that "away from" is convex.

My way to handle route prepositions is to view them as sequences of two conjoined paths: *by* or *past* a landmark means that the first path goes *near* the landmark, and the second path goes *away* from the landmark; *through* a landmark means that the first path goes *into* the landmark, and the second path goes *out of* the landmark; and going *behind* a landmark means that the first path goes *to behind* the landmark, and the second path goes *from behind* the landmark. Given these descriptions, the convexity of the route prepositions follows essentially from the convexity of the set of points that conjoin the two paths, in analogy with the convexity for the goal prepositions. A point conjoining two paths p and q is a point for which $p(1) = q(0)$. It is interesting to note that with the definition of betweenness for paths, conjoining paths preserves convexity: If P and Q are convex sets of paths and $P{\cdot}Q$ is the set of conjoinings of paths from P and Q, then $P{\cdot}Q$ is convex too.[13]

I finally turn to the convexity of "around." The corresponding sets of paths are not polarly convex. Take the "around" paths p_1 and p_2 that have the same radius $r = 1$ and the same starting angle $\alpha = 0°$ but go in opposite directions (clockwise versus counterclockwise). Consider now a path p_3 that also has a radius $r = 1$ and an angle that is always exactly halfway between the angles of p_1 and p_2. This path is not an around path. In fact, it is not even a proper path because it maps only to the angles $0°$ and $180°$.

A way out of this problem is to consider the *direction* of paths, so that the relation of betweenness is restricted to paths that have the same direction.[14] If we make this restriction, polar betweenness can also be defined for "around," and it can then be shown to be convex.

11.6   Domains of Prepositions

The second part of the single-domain thesis about prepositions concerns their domains. The traditional semantic approach to prepositions is that they express *spatial* relations. For example, Leibniz (1765, chap. 3, §1) writes that prepositions "are all taken from space, distance and movement, and then transferred to all sorts of changes, orders, sequences, differences." This *localist* view has then been a main trend in linguistics (Miller & Johnson-Laird, 1976; Jackendoff, 1983; Landau & Jackendoff, 1993; Herskovits, 1986; Zwarts & Winter, 2000). When combined with nonspatial words, they create a spatially structured mental representation of the expression.[15] For example, Herskovits (1986) presents an elaborate study of the fundamental spatial meanings of prepositions, and she argues that the spatial structure is transferred by metaphorical transformations to other contexts.

However, although the localist program has been successful for most locative and directional prepositions, recent analyses of prepositions have indicated that other domains than the visuospatial domain may be central for the meaning of some prepositions. First of all, there exist prepositions that refer to the *temporal* domain. In English, the clearest examples are "before" and "after." (The convexity of their temporal regions is immediate.) These words have etymologically a spatial origin but are now used primarily for the temporal domain. The spatial meanings have been taken over by "in front of" and "behind."

How can it be ascertained that the proper domain for "before" and "after" is the time dimension and not a spatial dimension? First note that all four prepositions are invariant under multiplication of the underlying dimension. There is, however, an asymmetry between "after" and "behind" in the following examples:

(11.6)   She is behind me in the queue, but if I turn around she is in front of me.

(11.7)   *She is after me in the queue, but if I turn around she is before me.

These examples show that a *reversal* of spatial orientation changes the preposition "behind" to its opposite "in front of." However, a spatial reversal does not change the temporal ordering from "before" to "after." This means that the meanings of "before" and "after" allow some spatial rotations, which is evidence that the prepositions are not based on the spatial domain.[16] As I will show, this argument concerning rotations can be generalized.

11.7   A Force-Dynamic Analysis of "In," "On," and "Against"

Several authors have proposed that the meanings of many prepositions include a *force-dynamic* component (Dewell, 1994; Bowerman, 1996a; Garrod, Ferrier, & Campbell, 1999; Tyler & Evans, 2001; Zwarts, 2010b; Beliën, 2002). I begin by considering the domains of "in" and "on."

I will use force vectors to analyze the dynamic aspects of these prepositions, although I cannot work out this idea in as much detail as the locative and directional prepositions, because the domain of forces is much more complex. Mathematically, what is needed here is the notion of a force field, that is, a space with a force vector associated with each point of that space. In most practical situations, such a force field would almost always involve gravitation as a component (this force has direction $\theta = 180°$). However, force fields may be extremely complex depending on the relations between the landmark and the trajectory, and it is therefore difficult to give a general analysis based on them.

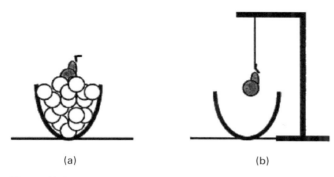

(a)   (b)

Figure 11.9
(a) The pear is in the bowl. (b) The pear is not in the bowl. Reprinted from S. C. Garrod, G. Ferrier, and S. Campbell. (1999). "In" and "on": Investigating the functional geometry of spatial prepositions. *Congnition, 72,* 167–189, © 1999, with permission from Elsevier.

Herskovits (1986) noted that the pear in figure 11.9a is considered to be *in* the bowl even though it is not spatially inside the bowl. If the apples are removed, but the pear is left in exactly the same spatial position as in figure 11.9b, then the pear is no longer *in* the bowl. So spatial location is not sufficient to determine whether an object is *in* a bowl. In figure 11.9a, the reason why the pear is *in* the bowl is that it is physically *supported* by the apples, while in 11.9b it has no such support. The notion of "support" clearly involves forces.

Vandeloise (1986, pp. 222–224) analyzes the French *dans* (in) in terms of containment, which he describes as a functional relation related to the notion of "carrier" (*porteur*) in that the container "controls" the position of the contained, but not conversely (p. 229). Similarly, the position of Garrod et al. (1999) is that if x is *in* y, then y's location "controls" x in the sense that the container y "constrains" the location of x (p. 173). They do not, however, specify what is meant by "constrains."[17] Also Zwarts (2010b, p. 209) writes that *in* "involves a passive and stative configuration of forces, not necessarily involving contact."

I propose that containment can be expressed as a counterfactual constraint based on *forces*: *if the trajector were perturbed, it would still be controlled by forces exerted by the landmark.*[18] This condition clearly separates the situation in figure 11.9a from that of 11.9b. I submit that this constraint captures the basic meaning of "in," and consequently the central meaning of the preposition is based on the force domain. In most cases, this can be replaced by a simpler condition: if the landmark moves, so does the trajector. Clearly this is not the whole story about "in" and containment because it does not distinguish it from "on" and support. Coventry, Carmichael, and Garrod (1994) and Feist and Gentner (1998) show that the concavity of the landmark plays a role. Subjects prefer "in" for dishes and "on" for plates.[19]

As mentioned earlier, a general analysis of convexity in terms of the forces involved in the meaning of "in" is extremely complicated. However, for situations of objects in containers, such as apples in a bowl, the force patterns involved would, in general, support the convexity of "in." I will not give a more detailed analysis.

Furthermore, because the forces controlling the figure have a spatial location, it is difficult to totally disentangle the force domain from the visuospatial domain and verify that only the force domain determines the meaning of "in." To be sure, there are examples where "in" is used purely spatially without any forces involved as in (11.8) and (11.9). One could view these as metonymic extensions of the central meaning of "in." In many of these cases, both "in" and the purely spatial "inside" can be used (e.g., "in the box," "inside the box"). There are, however, cases where only "inside" can be used:

(11.8)   The airplane is in the cloud.

(11.9)   Oscar is in the middle of the room.

(11.10)   inside the border, inside the city limits, inside the door

(11.11)   *in the border, *in the city limits, *in the door

In (11.10) "inside" seems to take a boundary as its landmark and refer to a region at one of the two sides of the boundary. In contrast to "in," then, "inside" is much more like the "axial" prepositions such as "behind" and "above" (Svenonius, 2006) and hence clearly a spatial preposition.

Another example is illustrated in figure 11.10. In figure 11.10a, the movements of the duck are controlled by the movements of the tube, and according to the proposed criterion, it is appropriate to describe the situation as that the duck is *in* the ring. In contrast, figure 11.10b shows a situation where the movements of the duck are less constrained by those of the tube, and consequently it is less appropriate to say that the duck is in the tube. It is more felicitous to say that the duck is *inside* or *within* the tube, since "inside" and "within" are prepositions that refer to the spatial domain.[20]

To measure the container's degree of control on the trajectory x, we can define the function $in(x)$ by saying that $in(x) = d$ if d is the average distance that the container is perturbed before x changes position.[21] In figure 11.10a,

(a) (b)

Figure 11.10
(a) A duck in a tube. (b) A duck not in a tube.

in(duck) = 0, since the duck moves as soon as the tube is perturbed. However, in figure 11.10b, *in*(duck) is the average distance of the duck from the sides of the tube, which is greater than zero.[22] Similarly, in Figure 11.9a, *in*(pear) = 0, but *in*(pear) = ∞ in figure 11.9b, if the container is imagined to move only in the horizontal dimensions.

For the preposition "on," the semantic representation involves contact and support from below. A spatial region is not sufficient to determine the meaning of "on." In brief, I propose that the meaning of "*x* is *on* y" is that *the force vector from x makes x come in contact with y, and a counterforce from y balances the force vector of x.* Typically, the force vector pointing from the figure *x* is generated by gravitation. (Just as with "in," the meaning of "on" also involves counterfactual control: *x* is *on* the table means that if the table were to move, so would *x*.)

How can it be established that the basic domain of "on" is the force domain? This is a problem because, in typical cases, the role of forces is not noted when "on" is applied. The reason is that the gravitational force is like the drone of bagpipe music: it is always there and normally not attended to. However, in other situations, the force dynamics are more transparent. Figure 11.11 illustrates a situation where the spatial relation involved in typical uses of "on" can be contrasted with the force-dynamic meaning of the preposition. The lamp is vertically above the balloon and in contact with it, which are the normal spatial conditions proposed for the meaning of "on." Nevertheless it is odd to say that the lamp is *on* the balloon. As a matter of fact, it might be more natural to say that the balloon is *on* the lamp or *against* the lamp, since the lifting force from the balloon makes it come into contact with the lamp (if the lamp moves,

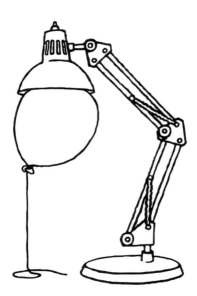

Figure 11.11
Is the lamp *on* the balloon?

so does the balloon, but not conversely). In this example, the directions of the forces involved are the opposite of what is typical.

The same principle applies to, for example, "the painting is on the wall," "the Band-Aid is on the leg," and "the button is on the shirt." What is crucial in examples of this type is that there is a force that acts on the trajector that makes it remain in contact with the landmark, not the spatial direction of the trajectory in relation to the landmark. In the case of the painting on the wall, it is still gravitation that acts on the trajector, but now the hanging mechanism makes the painting exert a pressure in the direction of the wall. These examples support the claim that only the force domain is necessary to determine the meaning of "on." More generally, this can be expressed by saying that "on" is invariant under spatial *rotation* as long as the force relations stay the same. As we have seen before, this provides evidence that the meaning of "on" depends on the force domain and not the spatial domain.[23]

Regarding the convexity of "on," the situation for the horizontal dimensions is relatively clear: if x and y are *on* an extended landmark u that is itself convex, and z is between x and y, then z is also *on* u, under the assumption that the force field operating on x and y toward u is convex, so that it also operates on z. However, there is also a "vertical" sense of convexity ("vertical" in relation to the direction of the force). For example, if a book is *on* a table and a plate is *on* the book, then one can say that the plate is also *on* the table. This form of transitivity is limited to horizontally extended objects, though: if an apple is *on* a table and the stem is *on* the apple, then we do not infer that the stem is *on* the table. Nevertheless, if *x* is *on* *u*, and *y*, which is above *x*, is also on *u*, then any object between *x* and *y* is also on *u*. In brief, "on" satisfies the convexity requirement if the relevant landmarks and force field have the required convexity properties.

Some authors speak of a *functional* analysis of prepositions rather than a force-dynamic one (Vandeloise, 1986; Coventry et al., 1994; Garrod et al., 1999; Coventry, Prata-Sala, & Richards, 2001; McIntyre, 2007). In line with the analysis of functional properties in section 8.5, I suggest that most of the functions used in these analyses can be reduced to forces or force patterns. For example, Garrod et al. (1999, pp. 173–174) defines the notion of a landmark *y functionally supporting* a trajector *x* as follows: "*y*'s location controls the location of *x* with respect to a unidirectional force (by default gravity) by virtue of some degree of contact between *x* and *y*." They next say that if *x* is on *y*, then *y* functionally supports *x*. In terms of forces, this can be expressed by saying that gravitation (or some other force) presses *x* toward *y*, and the friction between *x* and *y* makes *x* move whenever *y* moves. Their analysis is congruent with the one presented here, though they do not explicitly mention the force domain and its geometric structure as a separate domain.

"Against" is perhaps the clearest example of a preposition that is based on the force domain (Zwarts, 2010b):

(11.12)   Oscar bumped against the wall.

Zwarts (2010b, p. 194) notes that "against" "combines with verbs like *crash, lean, push, bang*, and *rest*, verbs that all involve forces." In the typical case, such as (11.12), the trajector follows a more or less *horizontal path* and exerts a force on a landmark. There are, however, also static uses of "against" where the path is reduced to a point (endpoint focus transformation):

(11.13)   The ladder leans against the house.

Furthermore, in this case too the direction can be changed by a rotation:

(11.14)   Standing on the wooden stairs, Oscar pressed his shoulders against the cellar flap, but he could not open it.

In (11.14) the direction of the force is vertical. Again, this suggests that "against" is invariant to rotational spatial transformations.

Bowerman and her colleagues (Bowerman & Pedersen, 1992; Bowerman, 1996; Bowerman & Choi, 2001) analyze *aan*, *op*, and *in* in Dutch, as well as the Korean prepositional verbs *nehta* (put loosely in or around) and *kkita* (fit tightly in). The analysis involves force components that cannot be reduced to spatial relations.[24] Bowerman and Choi (2001) present a semantic map containing five steps that is divided into different areas by different prepositions in different languages.[25]

11.8   "Over" as a Force Relation

Within the tradition of cognitive semantics, the preposition "over" has been studied over and over again, beginning with Brugman (1981), then expanded by Lakoff (1987) and partly reanalyzed by, for example, Vandeloise (1986), Dewell (1994), Kreitzer (1997), Tyler and Evans (2001), Zlatev (2003), and Beliën (2008).

Coventry et al. (2001) propose that what distinguishes "over/under" from the purely spatial "above/below" is that "over/under" has a functional meaning. They showed subjects, for example, pictures of people wearing umbrellas protecting them with more or less success against the rain. Their results suggest that the use of "above/below" is determined by spatial relations, while "over/under" is sensitive to functional relations, for example, whether the umbrella is protecting a person from rain falling in a slanted direction. I agree that Coventry et al. (2001) are on the right track, but instead I would argue that "over" has a central meaning that is based on a relation in the force domain (see also Coseriu, 2003; Van der Gucht et al., 2007; Beliën, 2008).

A common assumption in the kinds of lexical analyses performed in cognitive linguistics is that a word or an expression has a prototypical meaning, which can then be extended by different transformations. Brugman's (1981) and Lakoff's (1987) central image schema for "over" is depicted in figure 11.12. The content of the schema can be formulated entirely in terms of spatial dimensions as the trajector (TR) moving horizontally in a position vertically higher than a landmark (LM). A prototypical example is the following:

(11.15)   The bird flies over the yard.

On this account, "over" typically describes a kinematic scene (in contrast to "above" which is stative).

Starting from this central schema, Lakoff then identifies twenty-four different senses of "over" that are connected to each other in radial network. Most of these senses can be described as *elaborations* (Holmqvist,

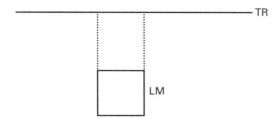

Figure 11.12
The central image schema for "over" according to Lakoff (1987, p. 426).

1993) or *superimpositions*, where the different elements of the schema are further specified. I do not count such elaborations as alternative meanings of "over"; they are just specializations of other meanings. Tyler and Evans (2001) are more systematic, and they present criteria for when two meanings of a word are different. Nevertheless they end up with fourteen different meanings of "over." Most of these can be generated from the central schema, just as in the case of Lakoff's meanings, by metaphorical and metonymical transformations.

Instead I want to argue in favor of a minimal specification of the meaning of "over." The crucial point concerning minimal specification versus full specification is that the semantics of a word must be separated from its role in a construal underlying a particular composite expression. Here I follow Fauconnier (1990, p. 400), who writes: "The 'semantics' of a language expression is the set of constraints it imposes on cognitive constructions; this is a structural property, which is independent of context" (for a similar position, see Coseriu, 2003). When several words are composed, their constraints are combined in the cognitively most efficient manner. If the constraints are incompatible, a metaphorical or metonymical transformation is required to close the meaning gap. The different kinds of combination of constraints will result in a variety of construals that may give the impression that a particular word is polysemous.

In contrast to most previous analyses, I therefore maintain that there is a single central meaning of "over" from which the other meanings are generated by various combinations and transformations. Dewell (1994) argues that the central meaning of "over" should be described as the trajector taking a *semicircular* path in relation to the landmark, as in figure 11.13.

The semicircular structure includes the feature that the trajector moves up and down along a path. In typical cases, the vertical dimension is associated with a gravitational force. Although Dewell only marginally considers the force relations, adding gravitation to the vertical axis entails

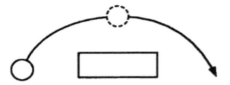

Figure 11.13
The central image schema for "over" according to Dewell (1994, p. 353). Reprinted with permission from De Gruyter.

that moving up along this dimension involves a force that is strong enough to overcome gravitation. I therefore propose that the central meaning of "over" is the schema in figure 11.13, but including a force-dynamic element in the description of the path.[26] In many cases, the countergravitational force is exerted by the trajector itself (e.g., "the bird flies over the yard"). Just as for "on," the basic domain for "over" is therefore the force domain. In contrast to "on," the central meaning of "over" involves a path and no contact between trajector and landmark. For "in" and "on," it is also assumed that the force of the trajectory is balanced by a counterforce from the landmark, but for "over" the forces of the landmark do not seem to play a role. Then, since force dynamics naturally lead to changes in spatial position, "over" generates a lot of implicatures for the visuospatial domain, but still, the force dynamics is primary.

Regarding the convexity of the meaning of "over," this again depends on how betweenness relations for force vectors and paths are defined. I will not go into the details of an analysis.

A similar force-dynamic analysis of Dutch *over* is presented by Beliën (2008). She describes the central constraint as follows: "*Over* designates a relation between a trajector and a landmark in which the trajector is related to the landmark by a mental path that follows a surface of the landmark, and from which a force points to the landmark" (p. 49). The central component here is that "over" involves a force that is directed toward the landmark. The situation is parallel to that of "on" and "in." Just as for "on" and "in," the primacy of the force domain can be exhibited by considering spatial *rotations* of the basic scheme.

In the schema in figure 11.13, the canonical direction for "over" is the vertical dimension, determined by gravitation. In some cases, one finds transformations of the canonical direction:

(11.16)   Victoria wears a veil over her face.

(11.17)   Victoria held her hands over her eyes.

In both (11.16) and (11.17), the landmark is the face, and its canonical direction is transformed into the vertical dimension. In example (11.16), it is still gravitation that acts on the trajector, but now the fastening of the veil makes it exert a pressure in the *horizontal* direction toward the face. And in example (11.17), it is Victoria herself who exerts a force of her hand in the horizontal direction. From this perspective on the forces involved, the veil and the hands are *over* the face.[27]

Other force directions can be involved as well:

(11.18)   The fly is crawling over the ceiling.

In this example, the ceiling is the landmark, upside down in relation to the normal ground, but again over the ceiling in relation to the forces exerted on the trajector, the fly. The role of gravitation is downplayed, since it is overcome by the forces that make the fly's feet stick to the ceiling.

(11.19)   Oscar nailed a board over the hole in the ceiling.

In this example, the nailing creates an upward force that makes the trajector (the board) be directed toward the landmark (the ceiling). So in examples (11.18) and (11.19) the prevalent force is directed vertically *upward*, which is yet another example of invariance under spatial rotation (further examples are found in Coventry et al., 2001). The upshot is that the invariance of "over" under visuospatial rotational transformations strongly suggests that the force domain is primary for the meaning of "over."

I next give some examples of the transformations involved in adapting the central meaning of "over" (fig. 11.13) to different construals. Dewell (1994, p. 355) points out that Lakoff's schema (fig. 11.12) can be seen as a special case of his where the central region is profiled. This example brings up a first type of semantic transformation: *profiling* a segment of a schema.[28] For example, in "The bird flew over the yard," the central part of the bird's flying path is attended to. In contrast, in "Sam fell over the cliff," the downward part of the trajector is put in focus; and in "The plane climbed high over the city," the focus is on the upward part of the trajector (p. 356). Note that this profiling mechanism is a special case of the inner attentional process discussed in section 1.3. The profiling is not arbitrary, however; Dewell argues that a constraint for "over" is that the profiled segments must include the peak point of the arc (p. 355).

A special case of a profiling transformation is *endpoint focus*. In "Sausalito is over the bridge" (Lakoff, 1987, p. 424), Sausalito is not moving vertically above the Golden Gate Bridge, but the speaker's inner gaze makes a fictive motion (see sec. 1.3) from San Francisco over the bridge to

Sausalito. Focusing on the endpoint is a *pars pro toto* metonymy. This case can be seen as a counterexample to Dewell's constraint that the peak point must be included. Admittedly, it engages another type of transformation.

The central meaning of "over" both involves a path (located vertically higher than the landmark) and requires that the trajector is not in contact with the landmark. In English, both these components can be modified by transformations:

(11.20)   The car drives over the bridge (contact).

(11.21)   The painting hangs over the fireplace (no path).[29]

However, the two transformations cannot be combined. For example, a painting that is over the fireplace but in contact with it will be said to be *on* the fireplace, since the scene then fulfills the requirements for the basic meaning of "on." Dewell (1994, p. 373) notes that when "over" involves contact, it is distinguished from "on" by involving a path (can be mental scanning) and by the trajector (or its path) covering the landmark (as in [11.20]).

Endpoint focus can be applied not only to spatial phenomena but also to *processes*. In "The play is over" (Lakoff, 1987, p. 439), the play is construed as an extended event that creates a path in time. If the end of the path is focused, the play has gone *over* this path. Thus this example has the same structure as "Sausalito is over the bridge," except that the relevant domain for the endpoint focus is time, not space. Note that the use of "over" in "The play is over" is, consequently, metaphorical, according to the account in section 2.6, since the horizontal dimension is changed to the time dimension. "Over again" involves repeating an event and traversing the path repeatedly.

Finally, a few words about "under." In many respects, this preposition behaves like a complement to "over," though the force-dynamic features of "under" are different in some situations (see also Coventry et al., 2001). In contrast to "below," which has its meaning in the spatial domain, "under" indicates some dynamic interaction between trajector and landmark. McIntyre (2007, p. 2) gives the following example:

(32)   I washed it under/*below the shower.

However, I will not pursue the details of the analysis.

11.9   Conclusion

The main thesis of this chapter is that prepositions represent convex sets of vectors or paths in a single domain. By using polar coordinates to define betweenness, I have shown that both locational and directional prepositions in general fulfill the convexity criterion.

Regarding the thesis about a single domain, my claim is that many prepositions, traditionally believed to express spatial relations, involve force dynamics in their central meanings. A still unresolved question is whether the core meanings of, for example, "on," "in," and "over" can be expressed in the force domain only or their basic semantics also requires the visuospatial domain. For example, "over" and "against" both involve paths in their core meanings, but this is a path that is associated with a force field.

One preposition that I have not considered is "at." It is related to the spatial "by" and "toward," but examples indicate that the *goal domain* is included in the meaning of "at." Herskovits (1986) mentions that sitting *at* a desk or washing *at* a sink involves more than just being close to the desk or the sink. In these cases, there is an intentional component in being *at* a place. Similarly, Landau and Jackendoff (1993, p. 231) point out that throwing a ball *toward* someone is different from throwing a ball *at* someone. In the latter case, the throwing has an intention of hitting.[30]

With this caveat concerning "at," my tentative conclusion is therefore that locative and directional prepositions are based on either the visuospatial domain, the time domain, or the force domain.[31] What complicates matters is that there are many metaphorical uses of prepositions.

Most important, the ordinary locative prepositions can be used in other domains, for example, "the temperature is *above* 20°," "Midsummer is *behind* us," "The color of our neighbors' house is *near* that of ours," "Victoria is working *toward* her goals." And, of course, prepositions based on the force domain can be used metaphorically for the visuospatial domain. Since the force vectors are also spatially located, this kind of metaphor is difficult to detect.

Since prepositions, in most languages, form a closed class, new meanings cannot be introduced easily by an additional word. So when an expression for a relation within a domain is required, in particular for domains different from the visuospatial domain or the force domain, the most similar prepositional meaning will have to be selected (cf. Tyler & Evans, 2001, p. 761). Often a metaphorical or metonymical transformation helps in closing the meaning gap. The context of the expression adds information too (cf. Tyler & Evans, 2001, p. 762).

Just as adjectives can be used to express result vectors (see sec. 10.8), so a preposition together with a noun phrase corresponds to a result vector: "The car is *inside the garage*" is represented by a stative (zero) result vector, and "The car drives *into the garage*" is represented by a result vector going from the outside to the inside region of the visuospatial domain (with respect to the garage). In line with this, Zwarts (2008, p.103) writes: "We can view stative verbs as the counterparts of locative prepositions. Stative verbs like sleep or shine and adjectives like red, wide and dark do not refer to trajectories in conceptual space, but to single points or positions." Similarly, directional prepositions have close semantic affinities with result verbs. A further similarity between verbs and prepositions is that prepositions also have *telicity* (Zwarts, 2005). Locative prepositions are always telic. The directional prepositions can be divided between telic, such as "into" and "across," and atelic, such as "toward" and "along" (p. 746).

Langacker (2008, pp. 112–117) also writes about the parallels between adjectives, adverbs, and prepositions. He notes that prepositions have both "adjectival" ("the dust under the bed") and "adverbial" uses ("it is hot under the bed"). He comments: "This overlap is one reason for thinking that the traditional categorization—where adjectives, adverbs and prepositions are viewed as mutually exclusive classes—is less than optimal" (p. 117). Further support for this argument comes from the fact that words such as "near," "worth," and "like" can be used both as adjectives and as prepositions, and it is difficult to determine which word class they belong to (Maling, 1983). A simplifying way of expressing the connections is that the meanings of directional and locative prepositions are about the visuospatial domain, most adjectives are about object category domains, and result verbs are about both types of domains.

The main topic of this chapter has been the use of geometric notions to describe the semantics of prepositions. Locative and directional prepositions are prime examples of how meaning is geometrically structured. I have also shown that locative and directional prepositions support the general idea that concepts can be represented by convex regions in the visuospatial domain. In contrast, the force domain is involved in the meaning of prepositions such as "in," "on," "over," "under," and "against." Since the force fields involved depend to a large extent on the nature of the trajector and the landmark and their relation, it is difficult to determine to what extent the meaning of these prepositions represents convex regions in the force domain or some product space. This topic merits further investigation.

12   A Cognitive Analysis of Word Classes

12.1   Semantic versus Syntactic Accounts of Word Classes

Words are not just units of meaning; they come in classes. As described in section 6.1, when you first learn about word classes in school, you are presented with a brief gloss of what the different kinds of words are used for: nouns for things, verbs for actions, adjectives for properties, prepositions for spatial relations, and so on. In academic linguistics, however, word classes are almost exclusively defined by *syntactic* criteria. In contrast, my position is that the syntactic markers have evolved as *effects* of the divisions of words into categories, not as causes.

One of the aims of this book is to take seriously the idea that there are cognitive and communicative constraints on how words are grouped into classes (for a similar program, see Langacker, 1991b).[1] In previous chapters, I have presented semantic accounts of nouns, adjectives, verbs, and prepositions based on different constructions in conceptual spaces.

In this chapter, I briefly discuss the semantics of some more word classes. I cannot provide a systematic analysis of all word classes, so I confine myself to selected cases. I do not claim that any simple mapping holds between word classes and structures in conceptual spaces. As an illustration, Dixon (2004, p. 2) writes: "A lexical root cannot be assigned to a word class on the basis of its meaning. If this were so, then 'hunger/ (be) hungry,' '(be) mother (of),' '(be) two,' and 'beauty/(be) beautiful' would relate to the same class in every language, which they do not." Dixon also points out that the concept of *needing to eat* is expressed with nouns, adjectives, or verbs in different languages and that "mother" and "father" are verbs in some American Indian languages. Nor do words (word roots) necessarily belong to particular word classes. An example from English is "round," which can be used as an adjective, noun, verb, adverb, and preposition.

The idea that word classes can be clearly identified is a myth. The traditional categorizations of words by linguists, from Sanskrit grammarians onward, is an idealization that fits reasonably well for Indo-European languages but may not fit at all when classifying words in other types of languages, as has been pointed out by Croft (2001), Dixon (2004), and others.

12.2   Event Structure and the Roles of Words

In chapter 9, I presented a cognitive theory of events that is an idealization of a causal structure. To recapitulate, an event consists of a *force* vector, typically generated by an *agent* that affects a *patient* and *results* in a *change* in some *properties* of the patient. Already among the different components of the representation of an event, one can see the roots of a mapping to word classes. The fundamental distinctions are between agent and patient, on the one hand, and force vector and result vector, on the other.

I have argued that agent and patient categories are represented linguistically by *nouns* (noun phrases), and force and change vectors are represented by *verbs* (verb phrases). Thus the model of events and the thesis about construals explain why these two linguistic categories are fundamental.[2] Then the basic expressions can be expanded into more elaborate constructions, generating noun phrases and verb phrases. The important thing to note, however, is that the general structure of a sentence as consisting of at least a noun phrase together with a verb phrase is a consequence of the model of events presented in this book. This is a central example of how the semantic model constrains the syntactic structures. In brief, a sentence typically expresses a construal of an event that focuses either on the cause (force vector) or on the effect (result vector). Elaborating the other elements of the model of events will have multifaceted consequences for the structure of word classes and for how they are combined.

Language is full of transformations between word classes: the shape property *round* (adjective) becomes the shape of a result vector (*go*) *round* (verb), which then becomes the abstract noun (*a*) *round*; the noun *saddle* becomes the verb *saddle*; the verb *climb* is transformed into the noun *climber*; and so on. Several kinds of semantic transformations are involved in these transitions between word classes, and I cannot do justice to all of them. Many of them can, however, be explained as *refocusings within the event schema*. For example, in an event involving a force vector of climbing, there must be an agent doing the climbing, so this individual can then be determined via the verb for the force vector. By analyzing the different thematic roles involved in an event schema for special kinds of force and result vectors, it will often be possible to derive the semantic content of a word that has been transformed from another word class.

After these general comments on the relations between event roles and word classes, I now turn to some remarks on a few other word classes.

12.3   Adverbs

It is a challenge to describe the class of adverbs in semantic terms. There are many kinds of adverbs, for example, adverbs of manner, adverbs of frequency, adverbs of time, adverbs of place, adverbs of certainty. To some extent, the adverbs form a leftover class: adjectives modify nouns; adverbs modify everything else.

To give but one example of how adverbs are classified, Parsons (1990) divides them into speech-act modifiers, sentence modifiers, subject-oriented modifiers, verb phrase modifiers, and a remainder class. I cannot provide an analysis of all kinds of adverbs. As a complement to the semantics of verbs presented in chapter 10, I will focus on adverbs modifying verbs (verb phrases).

In the semantic model of verbs presented in chapter 10, verbs refer to vectors. Vectors can vary in terms of dimension, orientation, and magnitude. Therefore adverbs that are modifiers of verbs should refer to change in these features. For example, in "I speak slowly," the adverb selects one of the several dimensions from the sound domain of *speak*. "I speak loudly" selects another.[3] In "I walked backward," the adverb refers to the orientation of my motion. Finally, in "He pushed the door softly," the magnitude of the force vector representing push is diminished by the adverb. When an action involves a pattern of forces, adverbs can modify the whole pattern by providing dynamic information, for example, "she walked limply," "he smiled wryly," or "she kicked aggressively." What is common to these examples is that the adverb *restricts* the regions associated with the meanings of the verbs. Similarly, in relation to a result verb describing a concatenation of changes (as in a path), an adverb can provide information about the form the path takes, for example, "she crossed the park crookedly." In brief, the function of adverbs modifying verbs parallels how adjectives modify nouns.[4]

As long as adverbs function as multipliers (diminishers or magnifiers) within a particular domain, the convexity principle can be defended. For example, if certain voice volumes v_1 and v_2 both count as speaking *loudly*, then any volume between v_1 and v_2 will also count as loudly. And for adverbs expressing the form of a path, the principles of path convexity that were presented in section 11.5 will apply. It is an open question whether the convexity principle can also be applied to other adverbs.

Next I turn to the domain specificity of adverbs. In section 7.5, I showed how the structure of the domains underlying adjectives to some extent determines which adverbs can be combined with the adjective. Similar arguments can be applied to adverbs modifying verbs. In particular, force vectors have dimensions that are not shared by patient spaces representing results. The magnitude of force is peculiar to the force domain only, and therefore adverbs expressing such magnitude, for example, "strongly," should apply to manner verbs (push strongly, hit strongly), but not to result verbs (*move strongly, *fall strongly, *break strongly). This principle functions well for manner verbs that are represented by a single force vector, for example, contact verbs. For manner verbs that are represented by patterns of forces, however, the situation is less transparent (?walk strongly, ?swim strongly). While force magnitude is peculiar to manner, direction clearly is not. Hence modifiers expressing direction are generally shared by force and result vectors, for example, "ahead" and "left" (push ahead, move ahead). On the other hand, some adverbs are tied to domains that are mainly associated with result verbs, and thus they cannot be combined with manner verbs. For example, "darkly" relates to the light or the color domain but does not apply to the force domain (glimmer darkly, *push darkly, *hit darkly).

These considerations give some support for a thesis that I formulate as a first approximation:

Thesis about adverbs: Verb-modifying adverbs refer to a *single domain*.

This thesis can be defended at least for adverbs functioning as multipliers. To what extent the thesis is more generally valid, also for other types of adverbs, remains to be further evaluated. One limitation is that adverbs modifying adjectives (and other adverbs) can be zero-dimensional, for example, "very" and "completely," and therefore cannot be associated with any domain.

12.4   Pronouns, in Particular Demonstratives

What is the meaning of pronouns such as "she," "his," and "that"? It seems impossible to determine, since the referent of a pronoun changes almost as often as it is used (Chierchia, 1995). In section 1.5.2, however, I made a distinction between fast and slow changes of meaning. The word classes that have been considered so far all have meanings that change slowly (although a word can sometimes be given a temporary meaning in a particular communicative context; see, e.g., Krauss and Glucksberg, 1977). What characterizes pronouns is that they are given meanings during the *fast* semantic process. Nevertheless the semantic mechanisms are of the same type as in the slow process. The domains of pronouns are in general object categories, often persons. As an example of how the semantic structures that I have presented for other word classes also apply to pronouns, I will consider demonstratives.

Diessel (2006) claims that demonstratives—"this," "that," "here," and "there" in English—form a word class of their own. He argues that their role in communication is to create or emphasize *joint attention* among the interlocutors. Demonstratives are among the first words learned by children, often the first noncontent words learned. In line with the arguments in chapter 4, the earliest demonstratives that children produce are accompanied by declarative pointing (E. Clark, 1978). The combination of pointing with a demonstrative provides the child with an efficient tool to refer to objects in the surroundings, though the child does not yet know any category word for the object. Diessel (2006, p. 481) claims that "no other class of linguistic expressions . . . is so closely associated with a particular type of gesture than demonstratives."

In the development of language, demonstratives are primarily used for reference to the outer world—the immediate environment. In English the demonstratives can be classified along two domains: (i) "this" and "that" refer to the object category space, while "here" and "there" refer to the visuospatial domain; (ii) "this" and "here" refer to close objects or locations, while "that" and "there" refer to more distant objects or locations. Other languages do not always have the same classification of demonstratives, but demonstratives can be found in all languages (Diessel, 2006).

It might be objected that "there" is an example of a word, the meaning of which is a nonconvex region: regions at some distance to the left of me are *there*, and so are regions at some distance to the right of me, but regions in between are *here*. Thus convexity seems to be violated. However, just as for the prepositions "near" and "far," the principle still holds if polar coordinates are used instead: anything below a certain (context-dependent) distance, independent of angle, is *here*; anything above that distance, independent of angle, is *there*.

Later in language development, demonstratives are also used to refer to elements of the joint inner world of the interlocutors—their *common ground* (H. Clark, 1996). Diessel (2006) calls this the *discourse* use of demonstratives. This (*sic*) is a form of pointing to the inner world that was discussed in section 4.5. Diessel distinguishes between two subtypes:

(i) An *anaphoric* use in which the demonstrative refers to a previous discourse object.

(12.1)   The organizers of the exhibition gave the mayor a T-shirt with their logo. The mayor then wore *this* T-shirt for the rest of the day.

(ii) A *discourse deictic* use in which demonstratives refer to statements.

(12.2)   Oscar and Victoria argued all evening. *That* is the reason they broke up from the party so early.

In the discourse deictic use, demonstratives are often used to establish links—causal or argumentative—between statements.

To these two uses, one can add *metonymic* uses of demonstratives (H. Clark, 1992, chap. 3). For example, if someone says, "I used to work for *those* people," pointing at a newspaper, the gesture is metonymically referring to the people who produce the newspaper.

12.5   Introducing Anycat: Quantifiers and Negation

Every lie creates a parallel world—the world in which it is true.
—Momus

The approach to semantics that is based on logic has focused on the meanings of closed word classes: connectives and quantifiers. The flagship of truth-functional semantic theories is their treatment of *quantifiers*—words like "all," "some," "any," and their kin. In first-order logic, quantifiers have their own symbols: $\forall$ is used for *all* and $\exists$ for *some*. The original extensional analysis of quantifiers in logic has been extended to intensional theories (possible-worlds semantics). In linguistics, *Montague grammar* is the prime example of how quantifiers (and some other intensional concepts) can be treated in a truth-functional semantics (Montague, 1974; Lewis, 1970).

However, a cognitively based semantics should likewise have something to say about the meaning of quantifiers and related words, though they may be far from the basic meanings related to perception and action. I am aware that I am entering a minefield here; hence I will only treat some basic cases of the meaning of quantifiers.

I will outline an analysis inspired by Langacker (2003), who shows that expressions like "any cat" and "some cat" function as special cases of concepts involving fictional objects (points in conceptual spaces). Before I present the analysis, note that traditional truth-functional semantics normally sweeps a number of features of quantifiers relating to *context* under the carpet. When somebody says, "Everybody is here," the reference of *everybody* is not all human beings but some conversationally or contextually determined class E of individuals. For the utterance to have its intended meaning, this class must be shared between the interlocutors. Thus the reference of a quantifier, just as for other pronouns, is determined during the fast process of meaning change.[5]

Langacker (2003) points out that the quantifiers in English can be divided into two basic kinds: *proportional* (all, most, no) and *representative instance* (every, each, any). "Some" is used in both ways.[6] Both types implicitly refer to a contextually given maximal class E of entities—the class that the quantifier is about.

The meanings of the proportional quantifiers are decided in relation to E. The construal involves making a *multiplex-mass interchange* (see sec. 1.3.3), so that no single element of E is attended to, but the proportions are estimated. Then the relevant masses are compared to the proportion of E that they cover. "All X are Y" means that 100 percent of the Xs in E are also Ys. "Most X are Y" means that at least 50 percent of the Xs in E are also Ys (pragmatically, *most* in general means at least two-thirds) (Solt, 2011). "Some X are Y" means that the proportion of the Xs in E that also are Ys is greater than zero. Finally, "No X are Y" means that the proportion of the Xs in E that also are Ys is zero. This analysis applies also to *almost all* (close to 100 percent) and *almost no* (close to 0 percent).

A seemingly curious fact about the representative-instance quantifiers "every," "each," and "any" is that they function grammatically as *singular*, though they are universal quantifiers concerning all members of E. For these quantifiers, the elements of E are not blurred into a mass but construed as individuals.[7] The cognitive mechanism is that the quantifier generates a construal where a single element of E, albeit fictive, is attended to (put in focus). As evidence of the fictivity of the selected element, Langacker (2003) points out that identifying questions do not make sense:

(12.3)   *A*: Every/each/any cat is lazy. *B*: *Which one?

The representative-instance quantifiers build on different fictive mental operations. Following Langacker (2003), the operations for the quantifiers are as follows:

• *Any*: Arbitrary selection of one of the members of *E*.[8]
• *Each*: Sequential examination, one at a time, of the members of *E*. This operation is normally applied only when *E* contains a limited class of entities.
• *Every*: Functions like *each*, except that the scanning is not sequential. Langacker (2003) writes that the operation "evokes the image of the members of *E* being simultaneously visible and available for examination."

The arbitrarily selected cat is an individual in a construal and therefore functions linguistically like an individual: *Anycat*. Anycat can, for example, be referred to by a pronoun: "If any cat ate the salmon, *it* will be sorry" (Langacker, 2003).

If Anycat is an individual, what properties does it have? Since a statement about Anycat should apply to all cats in *E*, by its arbitrary nature, it follows that Anycat can only have the properties that all cats in *E* have.[9]

Langacker notes that the referent of "any" is necessarily fictive, whereas the indefinite "a" can focus either an actual or a virtual entity. Thus "A/*Any friend called me yesterday." As I noted in section 5.3, the generic use of "a" functions like "all."

Finally, a few words on negation. Within the logical tradition, negation has been treated as an operator on sentences. The meaning of "not" has been analyzed as reversing the truth-value of a sentence. There are, however, many uses of "not" and other forms of negation that do no apply to sentences (e.g., verbs and adjectives can be negated: "hook" becomes "unhook," "tall" becomes "not tall"). To understand the semantics of negation, truth-functional analyses should therefore be amended.

Instead I turn to the communicative function of negation. On the first level (instruction), negation typically functions in *prohibitions*: "Do not open the window!" By using such an utterance, the speaker forbids the hearer to perform a certain action. On the second level (coordinating worlds), the typical role of negation is to *revise* a suggested (or expected) expansion of the common ground in a dialogue.[10] For example, if *A* says, "He is a good cook," *A* suggests that this content be added to the common ground, but if *B* replies, "No, he is not," *B* refuses the addition. Thus negation plays a significant role in the negotiation of the common ground that will be shared by the interlocutors. Verhagen (2005) analyzes this perspective at length. He writes that "the primary function of negation should be understood in terms of cognitive coordination, not in terms of the relation between language and the world, or the language user and the world"

(2005, p. 26). Verhagen and others also argue that negation functions as a "space builder" (Fauconnier, 1984; Langacker, 2003; Verhagen, 2005, pp. 29–30). When a negation of a sentence *s* is uttered, the participants in a conversation open up another mental space where the negation of *s* is accepted, which then becomes part of the shared world.[11] Therefore it can be said that the semantic function of negation, like pronouns, is part of the fast meaning process, in contrast to the perspective of the logical tradition. It is perhaps only on the third level of communication (meaning coordination) that negation has its standard logical role, typically in generic statements.

12.6   The General Single-Domain Thesis

In this chapter I have presented a somewhat rhapsodic analysis of some word classes apart from the nouns, adjectives, verbs, and prepositions that have been treated in previous chapters. The observant reader has noticed that, with the exception of quantifiers, I have tried to argue for a general semantic rule:

General single-domain thesis: Words in all content word classes, except for nouns, refer to a single domain.

I am not certain how far I can push this thesis. To a large extent, its validity depends on how abstract domains are described. There are also word classes I have not considered, for example, connectives, where the general single-domain thesis may not be valid or may not even apply. And I have noted that certain adverbs are zero-dimensional and thus not dependent on any domain.

Nevertheless I want to put forward the thesis as a strong heuristic that language learners (implicitly) apply when learning the meaning of a new word. A default rule that if a newly encountered word is not a noun, then its meaning depends only on a single domain, will simplify the problem for the learner (cf. Bloom, 2000, chap. 8). In general, the syntactic markers of a word indicate its role in an event construal. Thus the markers help identify the relevant domain for the word. Lupyan and Dale (2010, p. 8) make "the paradoxical prediction that morphological overspecification, while clearly difficult for adults, facilitates infant language acquisition." Mandler (2004, p. 281) argues along the same lines:

Many of the grammatical aspects of language seem impossibly abstract for the very young child to master. But when the concepts that underlie them are analyzed in terms of notions that children have already conceptualized, not only does the linguistic problem facing the child seem more tractable but also the types of errors that are made become more predictable. The invention of grammatical forms to express conceptual notions that are salient in a young child's conceptualization of events seems especially informative.

Since the relevant domain is often determined by the communicative context in which the word is uttered, applying the single-domain thesis will make identifying the new meaning much more efficient. Thus I propose that the general single-domain bias provides one of the fundamental reasons why humans can learn a language as quickly as they do.[12]

13 Compositionality

It is astonishing what language can do. With a few syllables it can express an incalculable number of thoughts, so that even a thought grasped by a terrestrial being for the very first time can be put into a form of words which will be understood by somebody to whom the thought is entirely new.
—Gottlob Frege

In the previous chapters, I have focused on purely lexical semantics, that is, what can be said about the meanings of single words. However, all real communication involves composing simple meaning components into larger structures. The topic of this chapter is how conceptual spaces can be used to describe mechanisms of meaning composition.

The Fregean tradition of semantics has emphasized the combinatorial nature of meaning composition, whereby the meaning of a composite expression is fully determined by the meaning of its components and the rules for composing its structure (Szabo, 2004). Meanings of words are dynamic, however. They depend on the situation in which they are uttered (the context), and on the other words they are combined with (the co-text).[1] A consequence of this position is that, in contrast to the Fregean tradition, the linguistic expression underspecifies for meaning (Tyler & Evans, 2001).

One central idea of this chapter is that composition is not just putting meanings together, but combining meanings often involves *transformations* of the meanings of the components. It will be apparent that the semantic transformations studied in section 11.1 and elsewhere in the book play a central role in this process.

13.1 Direct Composition

Research on language has addressed the composition of meanings by different strategies. For example, research on metaphor has introduced non-Fregean meaning composition as mappings across semantic domains (Holyoak & Thagard, 1996; Fauconnier & Turner, 1998; Gärdenfors, 2000). Furthermore, conversation analysis has emphasized a sequential composition of meaning through the accumulation of a common ground (Clark & Schaefer, 1989). In this chapter, I suggest ways to reconcile and unify these traditions. Again, the text will contain some mathematical concepts.

In the context of conceptual spaces, compositionality emerges directly from the framework of domains, product spaces, and functions that I presented earlier. Different word classes may involve different methods of composition. I focus here on adjective-noun combinations but also turn to noun-verb combinations at the end of the chapter.

As an introductory example of adjective-noun combinations, consider "blue rectangle." The meaning of this expression is defined as the Cartesian product of the *blue* region of color space and the *rectangle* region of shape space (which can be analyzed as in sec. 2.5.4). Note that the product of any compact and convex sets will again be a compact and convex set: thus the structural properties of conceptual spaces are preserved under this basic composition operator.

Not only are topological properties preserved but also the continuity of functions. If the functions $f\colon A \to X$ and $g\colon B \to Y$ are both continuous, then the product function $h = (f,g)\colon A \times B \to X \times Y$ is also continuous. The composition of continuous functions ($g{\cdot}f$) is likewise continuous. This allows one to concatenate functions while preserving their basic properties.[2] All taken together, the consequence is that compositionality can preserve, at a higher level of aggregation, the fixpoint properties of simpler meaning components that were presented in chapter 5. Compositionality can also help fine-tune the grain of the fixpoint approximation by allowing finer decompositions of a conceptual space. For example, a policeman helping a tourist find a certain building can progressively refine his description of the surrounding buildings by compositionally adding attributes— for example, "the yellow building with a large iron gate on the right of the post office"—making smaller and smaller the set of buildings under consideration. In this way, compositionality enriches the set of communication moves available to achieving a meeting of minds.

One can recursively create ever richer composite concepts that preserve the basic topological properties of the original conceptual spaces. This process seems to have no upper bound.

Going in the other direction, the blue rectangle conceptual region can be decomposed into its generating regions *blue* and *rectangle* via *projection*

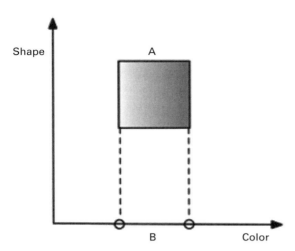

Figure 13.1
Projection on the color dimension.

(which in turn is a continuous function) from the product space to its component spaces (fig. 13.1).[3]

The analysis of meaning compositionality that I propose is not the same as the classical Fregean one. Compositionality is traditionally generated from the meaning of words or expressions, but in my account it is generated from domains and functions. Since each domain is associated with a class of words—for example, the class of color words—composing domains generates a product space, and the meanings of composite expressions can be located as regions of the product space.

An implicit assumption for the product space composition is that the domains of the product constructions are separable. The reality of linguistic usage shows that the spaces associated with composite expression are not fully independent; some *preprocessing* must take place before they can be properly composed. As an example, I will consider some cases of modifier-head composition.

13.2 Modifier-Head Composition

In the simplest cases—such as "blue rectangle," where "blue" is the *modifier* and "rectangle" is the *head*—the two associated domains (color and shape) can safely be assumed to be independent. However, this is rare in actual language use. More commonly, our knowledge of the space associated with the head may affect our representation of the modifier.[4] White wine is not white, a large squirrel is not a large animal, and a thick forest does not compare to thick hair. In all these cases, some preprocessing of the representation of the modifier space seems to be required to adapt it to one's knowledge of the head space.[5]

Some properties cannot be defined independently of other properties at all. Consider *tall*, which is connected to the height dimension but cannot be identified with a specific region in this dimension. To see the difficulty, consider that a Chihuahua is a dog, but a tall Chihuahua is not a tall dog. *Tall* presumes some *contrast class* given by some other property: things are tall not in themselves but only in relation to some other class of things. For a given contrast class Y—say, the class of dogs—the region $h(Y)$ of possible heights of objects in Y can be determined. A particular Y can then be said to be a tall Y if it belongs to the upper part of the region $h(Y)$.

For a contrast class such as skin color, one can map the empirically available colors on the color spindle (marked by stars in fig. 13.2). This

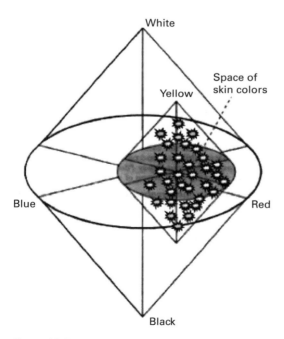

Figure 13.2
The subspace of skin colors embedded in the full color spindle (from Gärdenfors, 2000, p. 121).

mapping will determine a subregion of the full color space. If the subregion is embedded in a subspace with the same geometry as the full color space, one obtains something like figure 13.2. In the smaller spindle, the color words are used in the same way as in the full space, even if the hues of the color in the smaller space do not match the hues of that color in the complete space. Thus "white" is used to describe the lightest shades of skin, although "white" skin is actually pinkish; "black" refers to the darkest shades of skin, although most "black" skin is brown; and so on. And when someone is said to be "green with envy," this does not mean that there are shades of green in his or her face, only that the color belongs to the green area of the small color space, which corresponds to the "gray" part of the full color space.[6]

Provided that the head and modifier spaces are compact and convex regions of metric spaces, a way always exists of rescaling the distances in the modifier space to fit the constraints of the head space in a one-to-one correspondence. In this way, all color words will be available to characterize skin color. In general terms, *radial projection* (Berge, 1997) provides a natural conceptual tool to model such contextual rescaling effects. Consider two convex sets C and D defined within a space X and sharing an interior point 0 (taken as the *origo*). A radial projection is a mapping that establishes a one-to-one correspondence between the points in C and those in D, as well as between the border points of C and those of D. Such a correspondence redefines the distance between the points in C and the origo in terms of the distance between the corresponding points in D. Figure 13.3 shows an example of correspondence between points on the boundaries (x_0 and y_0) and points in the interior (x and y).[7]

This method can be formulated as a general principle: if the region of space representing the head contains a point shared with the space representing the modifier, this point can be taken as the origo of a transformation of the modifier space. The radial projection function tells one how to import structure from other domains. So long as concepts are convex and compact, such a function always exists.[8] Radial projection is a continuous function—so, again, all transformations preserve nearness.

If the head and modifier share only some domains, the modifier ("pet" in "pet fish") or the head ("lion" in "stone lion") is projected onto the shared subspace and then expanded into a shared space by inverse projection. With "stone lion," the representation of stone includes the property "nonliving," while "living" is presumed by many of the domains of "lion." These domains—for example, sound, habitat, behavior—cannot be assigned any region at all. By and large, the only domain of "lion" that is

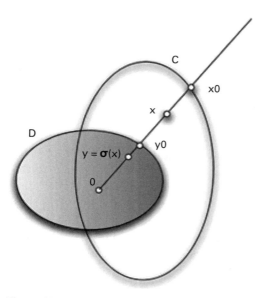

Figure 13.3
A radial projection.

compatible with stone is the shape domain. Consequently the meaning of "stone lion" is an object made of stone in the shape of a lion.[9]

13.3 Metaphorical Composition

Another class of head-modifier composition is based on metaphorical domain transformations (see sec. 2.6). Even if the head and the modifier do not share *any* dimensions, one can still create a mapping between the domains by exploiting their convexity and compactness. Indeed, an important consequence of radial projection is that any two convex, compact spaces can be mapped via homeomorphism. This permits the creation of metaphors as the transfer of structure from one domain to another. Once the transfer is made, one is back to the previous modifier composition.

As an example, consider the metaphor in "Our relationship has been bumpy." The literal meaning of "bumpy" refers to a structure in physical space, namely, an uneven (but continuous) distribution of values on the vertical dimension of a horizontally extended object, typically a road. This presumes two spatial dimensions, horizontal and vertical (fig. 13.4a).

A relationship between two persons is an abstract entity with no spatial location. How can a relationship be bumpy? My analysis is that the same

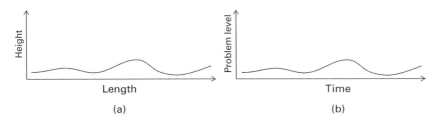

Figure 13.4
Depicting (a) a bumpy road; (b) a bumpy relationship.

geometric structure from the road is applied to a two-dimensional space: the horizontal spatial dimension is mapped onto the time dimension (the time interval of the relation), and the vertical spatial dimension is mapped onto a dimension of problem level (fig. 13.4b). The problem level dimension is typically conceived of as vertical: one talks about someone having a problem as an obstacle (see Lakoff & Johnson, 1980). Given these two mappings, the relationship is represented as a spatial curve that can be bumpy.

A metaphor does not come alone; it compares not only two concepts but also the structure of two complete domains. Once two domains have been connected via a metaphor, this connection may serve as a generator for new metaphors based on the same kind of relations (see also Lakoff & Johnson, 1980; Tourangeau & Sternberg, 1982; Gärdenfors, 2000, sec. 5.4). In brief, metaphorical mappings involve whole systems of concepts. For example, in section 3.4 I mentioned the metaphor of a computer virus. This is not an isolated metaphorical combination of words; rather, it invokes a mapping of several aspects from the domain of viral infections to computer programs. Thus the mapping can be used productively to generate new metaphors.

The metaphorical mappings highlight some of the dimensions of the source and the target domains. This can be an explanation of Fernández's (2007, p. 345) hypothesis that "metaphor interpretation involves enhancing properties of the metaphor vehicle that are relevant for interpretation, while suppressing those that are irrelevant." In other words, the metaphorical construction increases the prominence of the dimensions involved in the mapping.

I have now presented three ways of composing a modifier and a head. The first is plain compositional product construction. The second presupposes an adaptation of the modifier space, as a radial projection onto the head space. The third, involved in metaphor, requires a similar homeomorphic mapping between two disjoint spaces. All three cases presume that conceptual spaces are partitioned into domains. The topological properties of the spaces are preserved, preparing them for further composition.

The composition discussed in section 13.1 is really a special case of the composition in section 13.2, which in turn is a special case of the composition in this section. The first type need not modify existing spaces. The second modifies spaces that are naturally overlapping. Finally, the metaphors in this section require establishing homeomorphic correspondences between the disjoint spaces. A consequence of this analysis is that each of the three levels of composition should in turn require increasing cognitive processing. However, I do not know of any empirical studies that have tested this prediction.

Pragmatic considerations generate the position that meaning is largely underdetermined until context is brought in. The theory presented in this book suggests that context can affect the determination of meaning in two ways. The first is by being added to the common ground, as described in chapter 5. The second is through the modification of representations. I have already discussed three ways in which such modification may happen: (1) the contrast effect whereby vague terms like "tall" become more determinate; (2) the metaphorical mappings that allow introducing additional content, especially when the speaker has reason to believe that the addressee lacks compatible representations; and (3) the priming effects that make some dimensions more prominent.

Wisniewski (1996) studied concept combinations experimentally, in particular noun-noun combinations. He distinguishes between *property mapping*, where a property of the modifier is assigned to the head concept, and *relation linking*, where the combination is viewed as involving a relation between the two nouns. The examples I have discussed here are cases of property mapping. By relation linking, Wisniewski means cases such as "mountain stream" (stream *in* the mountains), "electric shock" (shock *caused* by electricity), and "honey bee" (bee that *makes* honey). Some of these relational compositions can be understood in terms of metonymic transformations, for example, zooming in or zooming out, that should be performed on the modifier before combining it with a head noun (see sec. 2.6). Others depend on functional properties that I described in section 8.4, and yet others on metaphorical mapping of the type described in this section. The upshot is that there exists a wealth of methods of modifier-head composition. As Wisniewski (1996) notes, we have no theory that can account for all the different forms found in natural communication. At its present stage, this is also true of my analysis, but hopefully the analysis of modifier-head composition can be expanded by bringing in further elements of the semantic model. For example, causal relational linking (e.g., "electric shock") should be analyzed by exploiting the thematic roles involved in event representations.

13.4 Noun-Verb Composition

Composing a noun with a verb can be regarded as the most fundamental form of word composition, since it generates sentences. In linguistics, verbs have been regarded as supplying the relational framework for expressing an event. Also in case grammar, the verb specifies how nouns relate to the event expressed. If the verb has this central role, it should be a highly stable element (cf. Gentner & France, 1988, p. 377).

If we look at the semantic uses of verbs in combinations with nouns, however, the meanings of a verb seem to vary a lot. For example, Keenan (1984, p. 201) notes that the meaning of "run" varies with the nature of the noun it combines with:

(13.1)   John/the horse is still running.

(13.2)   The car motor/my watch is still running.

(13.3)   The faucet/my nose is still running.

(13.4)   *My Fair Lady*/the Braque exhibition is still running.

In my analysis, the reason for such a variation in manner verbs is that a particular force pattern can be applied to many situations. The basic force pattern of "run," prototypically as exerted by a human, can be extended to similar movements for four legs, when combined with "horse," and to less-similar movements of mechanical parts, when combined with "motor" or "watch." Thus verbs generate many metaphors, for example, different meanings of "running." In a running faucet and nose, it is mainly the continuously repeated movement that is left, but the force pattern is generated by gravitation and no longer by the moving object. When a play or exhibition is running, it is also the continuous repetition of the event that is referred to, but here the intentions of people involved in the theater or the exhibition function as forces that "drive" the event. Another example is "cut" (Majid, Gullberg, van Staden, & Bowermann, 2007; see also examples 13.5–13.7 hereafter), where the result of the force vector depends to a large extent on the category of the patient, that is, the object being cut. As I argued in section 10.2, force vectors can also easily be used in a metaphorical way.[10]

Furthermore, Keenan (1984, pp. 202–203) argues that the meaning of the verb varies with the direct object of a sentence and with the subject of an intransitive predicate, but not with the subject of a transitive predicate. According to the semantics of verbs presented in chapter 10, this observation can be stated without relying on syntactic categories as follows:

Thesis about noun-verb combinations: The meaning of a verb is modified by the patient it combines with, but not by the agent.

The event model of chapter 9 explains the thesis, since both the force vector and the result vector of an event are related to the patient, and the outcome of the force vector depends on the nature of the patient. The thesis can be illustrated by the following examples (modified from Keenan, 1984, p. 20):

(13.5a)   Oscar cut his foot.

(13.5b)   The dog cut its foot.

(13.6a)   Oscar cut the lawn.

(13.6b)   The robot lawn mower cut the lawn.

(13.7a)   Oscar cut the rope.

(13.7b)   The sharp rock cut the rope.

The meaning of "cut" varies considerably between the three pairs (13.5), (13.6), and (13.7), although the force patterns involved are similar, which explains why they are all instances of "cut." However, the meaning does not vary much within each pair, supporting the thesis that the meaning of a verb is not modified by the agent of the event.

Keenan (1984, p. 203) also observes that many transitive verbs require special kinds of patients. For example, "peel" requires patients with a tegument, "spill" requires patients to be liquids or mass objects with relatively small granular size, and "shatter" requires patients that have special physical properties. In contrast, there seem to be no verbs that constrain the agents they combine with in a similar way.

These observations fit well with Gentner's (1981) "verb mutability" hypothesis, which proposes that the semantic structure conveyed by verbs is more likely to be altered to fit with other words than are the semantic structures conveyed by nouns or noun phrases (see also Gentner & France,

1988). For example, Gentner informally asked people about the meaning of "The flower kissed the rock." Most people kept the ordinary meaning of "flower" and "rock" but altered the meaning of "kiss" so that it could be performed by a flower, instead of changing the meanings of the nouns to flowerlike and rocklike people. The great flexibility of verbs can be explained for manner verbs, since it is only the force vector that has to be preserved and can easily be adapted to new noun contexts. However, the theory in chapter 10 also predicts that mutability should be more restricted for result verbs, since they are tied to domains with an added information structure. This prediction remains to be evaluated.

The analysis of noun-verb combination can be extended by considering the entire event model presented in chapter 9 and the thesis about event construals. The different components (thematic roles) of the model constrain each other in various ways. These constraints will have a large impact on which compositions of words are possible and what their meanings will be. The mechanisms I have presented here are just the beginnings of such an analysis of compositionality. What unites them are the underlying geometric and vectorial models.

III   ENVOI

14 Modeling Meanings in Robots and in the Semantic Web

Since my previous book on conceptual spaces (Gärdenfors, 2000) was published, a large number of applications of the theory have been developed, primarily within computer science, robotics, and information retrieval, but also in cognitive science, philosophy, linguistics, and psychology. From my own perspective, the most surprising developments have taken place within geoinformatics (e.g., Raubal, 2004; Ahlqvist, 2004; Schwering & Raubal, 2005; Adams & Raubal, 2009).

In this penultimate chapter, I discuss some potential computational applications of the semantic theory that has been presented in this book. In particular I want to point out the relevance of the theory for the development of the Semantic Web. Although there is still some distance from theory to computational implementation, I want to outline what types of new problems can be tackled on the basis of the theory.

Humans understand the meanings of words without being aware that they do so and without any knowledge of the underlying processes. If we turn to computer programs or robots, how could it be determined that they understand?[1] The two basic tests for artificial agents are that they can communicate and that they can draw inferences.[2] I turn first to understanding in robots. In the following sections, I then critically discuss the Semantic Web project and show how a semantic theory based on conceptual spaces can improve search mechanisms and question-answer systems so that they draw richer and more natural inferences.

14.1 Robot Semantics

In computer simulations and robotic experiments performed by Steels and others, the typical communicative situation is a *naming game* where the signaler, by uttering an expression, tries to make the recipient identify a particular object in the environment (e.g., Steels, 1999; Steels & Kaplan,

2002; Wellens & Loetzsch, 2012). Like the color naming by Jäger and van Rooij (2007) that I discussed in section 5.4.3 (see also Bleys, 2012), these games are good simulated illustrations of the process that generates a meeting of minds. A common feature of naming games (as in Wittgenstein's language games) is that the participants are only concerned with finding the appropriate referent among those that are *present* on the scene. The advantage of this is that the meanings that emerge automatically become *grounded* in the environment (Harnad, 1990; Williams et al., 2009).

In spite of all the successes of the simulations of naming games, it is not clear how to generalize these to communication about *nonpresent* referents. This kind of communication demands that the communicators have more advanced representational capacities. In particular, it would be interesting to see simulations where the robots *plan* to coordinate their future cooperation. This requires that the robots can represent actions and their consequences. Following the theory presented in this book, this means that they should have representations of events including the force and result vectors and the agent and patient properties.

The two-vector model of events has immediate consequences for how verbs can be learned and used by robots. Several groups have studied the topic of verb learning in robots (e.g., Cangelosi et al., 2008; Kalkan et al., forthcoming; Lallee, Madden, Hoen, & Dominey, 2010; Tikhanoff, Cangelosi, & Metta, 2011; Beyer, Camiano, & Griffiths, 2012). These attempts, however, focus on *result* verbs. For example, Kalkan et al. (forthcoming) used seven behavior categories: push-left, push-right, place-left, place-right, push-forward, place-forward, and lift. (Although "push" is a manner verb, it is used in the "move" meaning in this context.) Their algorithms for learning the verb meanings are based on "affordance relations" between entities, behaviors, and effects. In terms of the semantic model presented here, entities correspond to patients, behaviors to force vectors, and effects to result vectors. Kalkan et al. then present vector-based models for how an iCub robot can extract prototypical effect (result) vectors for the seven behavior categories.

I propose that this methodology can be extended also to manner verbs. For example, to be able to distinguish between "push" and "hit," the robot should calculate and categorize the force vectors in the actions. If it is observing another agent pushing or hitting, the force vectors can be extracted from the second derivatives of the kinematics of the movements (e.g., exploiting the methods of Giese, Thornton, & Edelman, 2008). As presented in chapter 7, the tradition from Johansson (1973) shows that the human brain is extremely efficient at this process.

Then the robot must learn the mapping from force vectors to result vectors. Hitting a ball will have different consequences from gently pushing it. This kind of associative mapping can be extracted from a combination of observing the force and result vectors of other agents interacting with objects and learning from the robots' own interactions with objects and their results (Cangelosi et al., 2008; Kalkan et al., forthcoming).

A complicating factor is that various contextual factors may influence the connection between the force and the result vector. For example, pushing an object on ice may lead to different results than pushing the same object in the same way on a lawn. This means that the learning involves a mapping from the three factors of force vector, object, and context to the result vector. Extracting a mapping of this kind is not an easy task, even in a simplified environment. However, by using clustering techniques or verbal input to the robot, the aim should be to learn a mapping from categories of force vectors and categories of objects to categories of result vectors that can also take some relevant contexts into account.

Implementing such a learning mechanism in an artificial system is a sizable task. However, this is the kind of mapping that children learn during their first years (van Dam, Rueschemeyer, & Bekkering, 2010). By manipulating objects in different ways and in different circumstances, children learn about the consequences of their actions. The learning is scaffolded by the language learning that is going on at the same time. I believe that the two-vector model of events that I have presented here is a powerful tool for implementing the learning of action meaning in artificial systems.

14.2 Why the Semantic Web Is Not Semantic

Given the enormous amount of information on the Internet, it is becoming ever more important to find methods for information integration. The Semantic Web is the best-known recent attempt in this direction. In an introductory article, Berners-Lee, Hendler, and Lassila (2001) write that "the Semantic Web is an extension of the current web in which information is given well-defined meaning, better enabling computers and people to work in cooperation." The ambition of the Semantic Web is excellent, but most work has been devoted to developing computer languages such as RDF for representing information and OWL for expressing ontologies. In my opinion, to enable "computers and people to work in cooperation," one should, above all, take into consideration how *humans* process concepts. As I argue in this chapter, this will be necessary if we want to put real semantic content into the Semantic Web.

The dream of the Semantic Web is to develop *one* ontology expressed in *one* language potentially covering everything that exists on the Web. Berners-Lee (1998) writes: "The Semantic Web is what we will get if we perform the same globalization process to Knowledge Representation that the Web initially did to Hypertext. We remove the centralized concepts of absolute truth, total knowledge, and total provability, and see what we can do with limited knowledge." As Noy and McGuinness (2001) note, there are several excellent reasons for developing ontologies: to share a common understanding of the structure of information among people; to enable reuse of domain knowledge; to make domain assumptions explicit; to separate domain knowledge from the operational knowledge; and to analyze domain knowledge. The question is whether the ontologies as we know them from the current Semantic Web are the best tools to achieve these goals. The theory of this book shows that there is much more to the semantics of concepts.

In practice, the picture is not so beautiful: there are several ontologies in several languages covering partly overlapping subdomains of the Web. And the formalisms encounter several kinds of integration problems, including structural heterogeneity, semantic heterogeneity, inconsistency, and redundancy problems (Visser, 2004).

Aiming for explicit description is one of the dogmas of the tradition of symbolic knowledge representations. I challenge the view that explicit symbolic characterization of ontologies is the most efficient, let alone the most natural way (in relation to human cognition) to represent this kind of knowledge for use on the Semantic Web. Shirky (2003) formulates the problem in the following way: "The Semantic Web takes for granted that many important aspects of the world can be specified in an unambiguous and universally agreed-on fashion, then spends a great deal of time talking about the ideal XML formats for those descriptions. This puts the stress on the wrong part of the problem—if the world were easy to describe, you could do it in Sanskrit." The main purpose of this chapter is to propose an alternative methodology and an alternative representational format for the Semantic Web based on conceptual spaces.

It is an unfortunate dogma of computer science in general, and the Semantic Web in particular, that all semantic contents are reducible to first-order logic or to set theory. Berners-Lee et al. (2001) claim that "fortunately, a large majority of the information we want to express is along the lines of 'a hex-head bolt is a type of machine bolt.'" Unfortunately this is not true. If one considers how humans handle concepts, the class relation structures of the Semantic Web capture only a minute part of our information about concepts. In particular, we often categorize objects according to the *similarity* between the objects (Goldstone, 1994; Gärdenfors, 2000). And similarity is not a notion that can be expressed in a natural way in a Web ontology language.

Along the same lines, Shirky (2003) declares that "the Semantic Web is a machine for creating syllogisms." He concludes: "This is the promise of the Semantic Web—it will improve all the areas of your life where you currently use syllogisms. Which is to say, almost nowhere." He adds, somewhat sarcastically: "The people working on the Semantic Web greatly overestimate the value of deductive reasoning (a persistent theme in Artificial Intelligence projects generally). The great popularizer of this error was Arthur Conan Doyle, whose Sherlock Holmes stories have done more damage to people's understanding of human intelligence than anyone other than René Descartes."

In Gärdenfors (2000), I contrast three basic methodologies within the cognitive sciences for representing information: the symbolic, the associationist, and the conceptual. In the symbolic approach, cognition is seen as essentially being *computation* involving symbol manipulation. The second approach is associationism, where *associations* between different kinds of information elements carry the main burden of representation. Connectionism is a special case of associationism that models associations using artificial neuron networks. The Semantic Web builds almost entirely on the symbolic methodology. My point is that if we want to build real content into an artificial system, we should rely on conceptual spaces. I claim that a Semantic Web worthy of its name will not get off the ground until the similarity aspects of concepts are accounted for. In conceptual spaces, similarity is modeled as distances, which is a computationally amenable way of handling the iconicity of concepts. Another important criterion for a successful computational model of the meanings of concepts is that it should be able to handle compositions of concepts.

14.3   Conceptual Spaces as a Tool for the Semantic Web

Conceptual spaces can be seen as a representational level that serves as an anchoring mechanism between language and reality. Via representations of sensory dimensions, many domains of a conceptual space are grounded in the "real world" (Harnad, 1990; Williams et al., 2009). As explained in Gärdenfors (2000) and in chapter 2 of this book, other domains can then be added by metaphorical and configurational extensions. The connection between a conceptual space and a symbolic description is obtained by mapping name symbols onto *points* in the space and the concept (property and relation) symbols onto *regions* of the space. In this way, the symbolic expressions are anchored in a conceptual space, which in turn is partly "grounded" in reality (see, e.g., Chella, Frixione, & Gaglio, 2001).

I next want to outline how the representational format of conceptual spaces can be exploited to generate the kind of semantic structures that are needed for a Semantic Web that is appropriate for human users (for further details, see Gärdenfors, 2004b).

The most important benefit of putting conceptual spaces at a representational level "below" the symbolic level is that the *concept hierarchies* (taxonomies) that are required for the symbolic structures can be generated almost for free from the conceptual spaces. The validity of a generic such as "a robin is a bird" will *emerge* from the fact that the region of the conceptual space representing robins is a subregion of the one representing birds (see sec. 6.4.2). In brief, if category prototypes and a metric domain are specified, a Voronoi tessellation can be computed that is sufficient to generate what is needed for an ontology in the sense specified earlier (Noy & McGuinness, 2001).

Identities of concepts are also immediate, once the mapping between concept symbols (predicates) and regions is established. For example, "vixens are identical with female foxes" will follow from the fact that "vixens" and "female foxes" are represented by the same region of the conceptual space.

Identities of names are handled in a similar way. Each name is mapped onto a point (vector) in the space. Thus if two names are mapped onto the same point, they are identical in meaning. Matters become more complicated if names are mapped onto partial vectors, where the values for some dimensions may be missing.[3]

Property characteristics such as transitivity and symmetry also emerge from the representational format of the conceptual spaces. For example, that comparative relations like "earlier than" are *transitive* follows from the linear structure of the time dimension and is thus an intrinsic feature of this relation. Similarly, it follows that everything that is green is colored, since *green* refers to a region of the color domain. Further, a property constraint such as that nothing is both red and green (all over) is immediate in the color domain of the conceptual space, since these words refer to disjoint regions of the color domain.

The upshot is that these aspects of ontologies of the type used for the Semantic Web emerge from the domain structures of the conceptual space. To put it bluntly, once the conceptual structure of an informational system has been specified, the semantic information that OWL and other metadata languages are supposed to add becomes more or less redundant.

Another important aspect is that once this domain structure has been specified to allow representations of concept inclusion, property identity, and name identity in the way I have outlined, then there is no need for an inference engine on the symbolic level. The job of a symbolic reasoning mechanism is taken over by (nonsymbolic) calculations on the conceptual level. These calculations are of a completely different nature from what is standard for symbolic inference engines.[4]

However, more than taxonomies of concepts can be used to generate inferences. The great strength of using conceptual spaces is that they can be used for representing many other aspects of semantics as well. In particular, the information about *similarity* that comes with the conceptual spaces can be used for various forms of approximate reasoning on the Semantic Web. For example, if I am searching on the Web for a particular wine that I want to buy and the wine is not available, then, with the aid of the similarity information provided by the domain structures of a conceptual space, I can ask for a wine that is similar to the one I am looking for. Since similarity depends on which dimensions are prominent, the system should ask me for what dimensions of the wine are most important before it determines which wines are the most similar. If one sticks to the symbolic representational format for the Semantic Web, this kind of reasoning about similarities is well nigh impossible (unless the dimensional structure is built in to the symbolic representation).

Another benefit of using conceptual spaces is that one obtains a much richer way of representing concept combinations, as explained in the previous chapter. Earlier in the book, I also argued that *metaphors* and *metonymies*, to a large extent, can be analyzed using the conceptual framework. These semantic mechanisms are long-standing enigmas for the symbolic tradition.

The program I propose involves a radical shift in focus on what metadata should be added to the Web. In the current Semantic Web, the information mainly concerns taxonomies and inference rules. If conceptual spaces are used as a foundational methodology, the focus will be on describing *domain structures*. This involves, above all, specifying the geometric and topological structure of the domains. Describing domains will require a different programming methodology from what exists in OWL and similar languages. The second step is to give information about how the resulting space is partitioned into concepts. I have shown that by using prototypes and Voronoi tessellations, this can be done in a computationally efficient way. Representing information by conceptual spaces requires computations that involve vectors, using inferences based on similarities, rather than inference mechanisms based tree searching in a rule-based symbolic approach.

Of course, the proposed methodology for generating semantic content is not without problems. Since so much is delegated to the structure of the domains, the questions of how to identify and describe domains become central. Will we not end up with the same problems as for symbolic ontologies? How should one, for instance, handle the situation if there are *competing* domain descriptions in two Web applications? The beginning of an answer is that there is in general more agreement on how to define a domain than on how to define an ontology for the existing Semantic Web. Even if we are faced with competing domains, we can compare their structure via their geometric and topological properties (just as in metaphorical mappings).

The main factor preventing a rapid advancement of different applications of conceptual spaces is the lack of knowledge about the relevant domains and the structure of their dimensions. Only for a few domains do we have sufficient psychophysical evidence to say that we have identified their geometric structure. For many of the domains that will be used in applications of the Semantic Web, this kind of information will not be directly available, but the programmer must rely on other methods to obtain the required domain structure. Here tools like multidimensional scaling or principal component analysis will become useful (see Gärdenfors, 2000, secs. 2.6, 7.1.4).

14.4   Conceptual Space Markup Language

Researchers have put forth several proposals for formalizing the theory of conceptual space in a way that could be turned into computer programs. An early proposal comes from Fischer Nilsson (1999), who presents a *conceptual space logic* in the form of an extended conceptual lattice formalism. Guizzardi (2005) extensively compares traditional ontological systems with conceptual spaces and proposes methods for integrating the two approaches.

The proposal that comes closest to the theory of this book is presented by Adams and Raubal (2009). They have developed what they call Con-

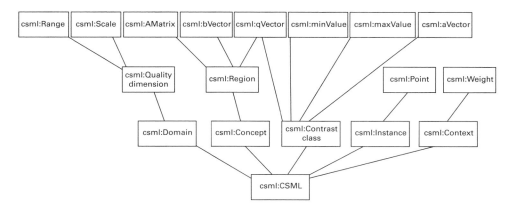

Figure 14.1
CSML tag hierarchy. © 2009 IEEE. Reprinted, with permission, from B. Adams and M. Raubal. (2009). The Conceptual Space Markup Language (CSML): Towards the Cognitive Semantic Web. In *Third IEEE International Conference on Semantic Computing* (pp. 253–260). Berkeley, CA: IEEE Computer Society.

ceptual Space Markup Language (CSML), an XML-based interchange format for sharing conceptual spaces so that they can be used in various semantic software programs. The core of CSML is a tagging hierarchy, where Domain, Concept, Contrast Class, Instance, and Context belong to the basic level (fig. 14.1). An example of the role of context is which rivers are judged to be *long*: A long British river is a short North American river. Domains are in turn specified by Quality Dimensions, Context by Weight (assigned to dimension), and so on. Knowledge bases written in CSML are stored as Web documents in the same way as ontologies written in OWL.[5]

As an application of CSML, Adams and Raubal (2009) show how a set of concepts from geoinformatics can be expressed in CSML. Figure 2.3 illustrated how the concepts *mountain* and *hill* can be represented as configurational regions (cf. sec. 2.5) in a space spanned by the width and height dimensions.

In comparison to other ontology markup languages, an important advantage of CSML is that it affords a method for measuring semantic similarity and richer methods for handling concept combinations (along the lines of the methods presented in chap. 13).[6]

CSML is just the beginning of a promising development. It focuses on representing objects. If it is to be useful also for more general question-answering systems or for robot-human interaction, it should be extended to cover actions and events as well. I hope the theory presented in this book will help in that task.

15 Taking Stock

15.1 Main Achievements

It is time to take stock. This book has presented a cognitive theory of semantics that fulfills all the criteria of chapter 1 in a constructive manner.[1] It has had two main goals: first, to introduce a theory of semantics as meetings of minds that combines cognitive linguistics with language games and conceptual spaces. Second, to show how geometric structures, including vectors, can be used to analyze the meanings of words in a way that grounds them in general cognitive mechanisms. I have argued that meanings are structured in domains. The structure provided by domains can help explain why children learn language so efficiently. The semantic analyses have been based on the communicative functions of various expressions rather than on their syntactic properties. The three levels of communication presented in chapter 5 have proved useful in separating out the basic semantic functions of various word classes.

An underlying assumption has been that most meaning structures exist in our minds independently of language. Consequently I have tried to keep the semantic theory free from influences from syntax. Unlike most other semantic theories in linguistics, my theory breaks out of symbolic representations of lexical roots and grounds their meanings in mechanisms based on perception and action.

I have cast my nets widely. The semantic theory that has been presented weaves together many strands of meaning: concept formation, language learning, sharing meaning, communicative function, lexical semantics, metaphors, compositionality, and so on.

Here is a summary of some of the achievements in the book:

• An analysis of *domains* based on conceptual spaces

I have argued that it is important to distinguish between meronomic structures and dimensional spaces. This results in a clear division of metaphors and metonymies, where metaphors refer to mappings between domains, and metonymies refer to meronomic and other relations within domains.

I have also highlighted the role of domains in language learning. In addition to perceptual domains, the force and the goal domains are central to my analysis. Words are not learned in isolation, but once a new domain becomes cognitively represented, the words that refer to different regions of the domain all become accessible. I presented a testable establishment hypothesis: if one word from a domain is learned during a certain establishment period, then other (common) words from the same domain are learned during roughly the same period.

• A model of meaning created by the meeting of minds

Linguistic meanings exist neither in the world nor in the head of a single individual. Semantics *emerge* as a result of the communicative interactions between the members of a community and their interactions with the world.

I have identified three levels of communication: instruction, coordination of common ground, and coordination of meaning. I have also introduced a distinction between fast and slow processes of meaning change. The fast process—for example, involved in the semantics of pronouns and many uses of quantifiers—has, in my opinion, been underestimated in previous research. As an example of the fast meaning process, I have analyzed the semantics and pragmatics of different kinds of pointing.

• A cognitive/communicative grounding of the main word classes

Following the lead of my previous book (Gärdenfors, 2000), I have emphasized the role of convexity in representing meaning regions in conceptual space. The convexity of concepts proves to be a strong general principle for cognition that can help explain a number of semantic phenomena. I have argued that the convexity of concepts facilitates learning and communication.

I have expanded the previous analysis of adjectives referring to convex regions in a single domain and nouns referring to convex regions in several domains. I have defended the thesis that the meaning of adjectives can be represented as convex regions in a single domain.

• A cognitive theory of actions and events

Actions have been grounded in the force domain. I have analyzed them as concepts in the configurational domain of force patterns. In parallel with object categories, I have proposed that an action concept is represented as a convex region in the domain of force patterns.

One of the major achievements of the book has been to present a cognitive theory of events that builds on conceptual spaces and the analysis of actions. The key idea is that an event consists of a force vector, a patient, a result vector, and often an agent. The force vector represents the cause of the event, often an action, and the result vector its effect. Other thematic roles can also be elements of an event. Another important notion is that of a construal, which consists of a particular attentional focus on some of the elements of an event.

• A semantic theory of verbs

The basic thesis is that a verb root refers either to the force vector or to the result vector of an event (but not both). From this it follows that verbs refer to a single domain: either the force domain, or the visuospatial domain, or an object category domain.

These theses are rich in predictions. I have shown that they can explain a number of semantic aspects of verbs, for example, similarity of meaning and subcategorization. They also explain the division between manner verbs and result verbs.

• A semantic theory of prepositions

I have shown that prepositions represent convex sets of points or paths in a single domain. My analysis is based on representing space in terms of polar coordinates rather than Cartesian. However, prepositions do not just concern the visuospatial domain; some of the common ones are based on the force domain, and some on the time domain.

In summary, part 2 of the book has shown that the convexity of concepts and the single-domain hypothesis for adjectives, verbs, and prepositions can explain a wealth of semantic data. They also generate new predictions that remain to be tested.

• A geometric theory of compositionality

I have shown how compositions of domains, using some mathematical transformations, can account in a general way for various forms of word compositions, among others, metaphors. This analysis of how meanings are composed can be applied to the endeavors concerning the Semantic

Web. It will generate a richer model of meaning than traditional methods that are based on logical techniques.

15.2   What Is Missing in the Edifice?

My aim has been to show that concepts from cognitive science, in particular conceptual spaces, can be used to model lexical semantics. As such, it is an attempt to establish a strong bridge between cognitive science and linguistics.

The goal to present the overall picture of a general semantic theory means that the mesh of my net must be coarse. For many areas within the lexical analyses, I have merely sketched a solution. Linguists may complain that the text contains too few examples. I agree. My defense is that this is meant to be a book, not a multivolume enterprise. I am well aware that each chapter deserves a book of its own.

The emphasis has been on presenting the general model and testing its viability. For this reason, I have not considered the details of the semantics of single words, as would be a common strategy for linguists. Of course, this does not exempt the model from rigorous testing against linguistic data. In this way, the book is a call for cooperation between linguists and cognitive scientists.

No theory is complete. Most of my examples have been rather concrete. I have focused on concepts that are directly grounded in perception and action, and left the analysis of more abstract concepts for future work. For one thing, it is not clear what domains are applicable to abstract concepts and how their structure can be determined.

My analysis of compositionality has presented some general principles. However, if future research considers constraints on the different elements of event construals, more specific forms of meaning composition can be analyzed. Furthermore, I have not considered the role of sentence connectives. I do not believe that they only have the logical function that has been in focus in much of the philosophical analysis, but believe that the meaning and function of the connectives also depend on discourse information and the structure of events.

The largest semantic unit that has been considered in this book is the event schema. However, events do not come in isolation. In communication, we combine construals of sequences of events into *narratives*. A topic for future investigation is how the structure of events in terms of manner and result vectors (and other roles) generates constraints on the structure of narratives. In particular, the causal structure of events should be respected.

15.3 Conclusion

Attempting to present a general theory of semantics involves a claim of unifying a number of research fields. I have striven to formulate testable theses in several areas of lexical semantics. As in all sciences, predictions run a great risk of turning out false or being considerably limited. Nevertheless, following the Popperian methodology, I find it more rewarding to be fruitfully wrong than to be boringly correct. My purpose in formulating the theses about lexical semantics is, above all, to stimulate further research on the topics treated in this book. I am eager to learn about the reactions of other researchers in semantics.

Appendix: Existence of Fixpoints

In this appendix, I present the mathematical background to the fixpoint theorem.[1]

Recall that a metric space is defined by a set of points with a measure of the distance between the points. A *metrizable* topological space is any space whose topological structure is imposed by some metric. My fundamental assumptions are that conceptual spaces are metrizable, and their metric structure is imposed by a similarity relation. This leaves open the possibility of many different metric structures. While the precise nature of psychologically sound similarity metrics remains highly controversial (and presumably differs between domains), numerous studies (e.g., Shepard, 1987; Nosofsky, 1988) suggest it to be a continuous function of Euclidean distance within conceptual spaces. However, the general ideas may be applied to other metric structures (see, e.g., Johannesson, 2002).[2]

Concepts are regions in a conceptual space. Two properties of such regions are worth mentioning. First, so long as concepts are closed and bounded regions in Euclidean conceptual spaces, they acquire a further critical topological property: compactness.[3] One important intuition regarding compactness is that it provides "enough" points that are near to a set; this proves critical when fixpoints come to be defined. Compactness makes it possible to approximate the entire space through a finite number of points, which also turns out to be critical.

In pointing as a way to achieve joint attention (see chap. 3), as well as in Jäger and van Rooij's (2007) model presented in section 5.4.3, it was assumed that the communicators share more or less the same conceptual space: in the pointing examples, the visual field; in Jäger and van Rooij's model, the color space. In general, however, different individuals will have different mental spaces. For simplicity's sake, assume that there are only two individuals A and B with spaces C_1 and C_2, both of which are convex and compact. If A communicates with B, A alters B's state of mind, and B's reaction will, in turn, change A's state of mind.[4] In more formal terms, communication can be described with the help of *semantic reaction functions* in the product space $C = C_1 \times C_2$. In my framework, the semantic reaction function represents the inherent bidirectional coupling of production and comprehension processes, a point common to much recent research on pragmatics and radical semantics (Dekker & van Rooij, 2000; Pickering & Garrod, 2004; van Benthem, 2008; Parikh, 2010). I assume that the reaction functions are continuous; that is, small changes in the communication will result in small changes in the reaction.

Now that all the ingredients are in place, I can simply remind readers of one of the most fundamental results of analysis: Brouwer's (1910) theorem, whereby each continuous mapping of a convex, compact set onto itself has at least one fixpoint. The continuous map I am concerned with in the present context is the semantic reaction function mapping the product space C onto itself. Note that C can be the product of several individual spaces and not just two. Figure 16.1a illustrates the fixpoint theorem for a function mapping a one-dimensional space onto itself, and figure 16.1b shows the necessity of the continuity assumption. The fixpoint is located where the curve crosses the diagonal.

What Brouwer's theorem says is that no matter what is the content of individual conceptual representations, as long as such representations are

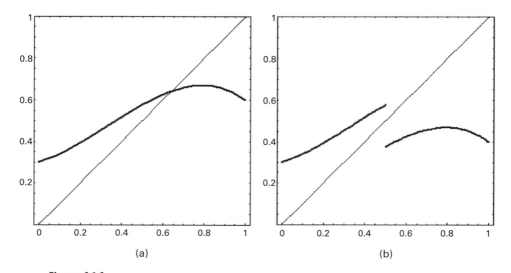

Figure 16.1
(a) The fixpoint illustrated for a one-dimensional space. (b) Fixpoints may not exist if the function is not continuous.

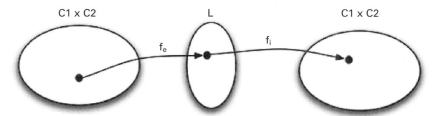

Figure 16.2
A semantic reaction function maps points in $C_1 \times C_2$ to $C_1 \times C_2$ via L, where f_e is the expression function, and f_i is the interpretation function.

well shaped and the communicative device is plastic enough to preserve the spatial structure of concepts, there will always be at least one point representing a meeting of minds.

Brouwer's theorem does not explicitly involve the role of language in how fixpoints are achieved. Since people are not telepathic, the mapping between individual conceptual spaces must somehow be mediated. Language is the primary mediator, although gestures and other such tools can also be used. Using language, the speaker maps his conceptual space onto some expressions from a language L, and the hearer in turn maps these expressions onto her conceptual space. Linguistic communication between two individuals with a product space $C_1 \times C_2$ is the composition of a function from C_1 to L and another function from L to C_2. This composition results in a modification of C_2, that is, a change of the hearer's mind (fig. 16.2).

The communicational intentions of other individuals can sometimes be rationally anticipated, leading to a kind of instantaneous adjustment of the interlocutors to a fixpoint, as in a standard communication game. In general, though, communication tends to be more myopic, leading to a sequence of partial adjustments that appear to be the rule rather than the exception in dialogue-based communication (H. Clark, 1992; Pickering & Garrod, 2004). As an example of an elementary adaptive process, consider once more the case of pointing, and let C_1 and C_2 be the visual spaces of two individuals A and B. For simplicity's sake, assume that $C_1 = C_2$. In the simplest case, the semantic reaction function starts from (x,y), where x is the point to which A is pointing, and y is the current position of A, which B is attending to so as to see the pointing direction of A. The reaction function maps the initial (x,y) to (x,y'), (x,y'') . . . (x,x), meaning that B is following A's gaze to the point x. The resulting fixpoint is the one at which A and B attend to the same object.

Brouwer's theorem proves the existence of fixpoints for any continuous function mapping a compact and convex space onto itself. However, the continuity of the semantic reaction function may seem too demanding an assumption in light of the inherent discreteness of the lexicon. The challenge of reconciling lexical discreteness and, more generally, the discrete symbolic nature of language with the continuous, spatial nature of conceptual systems has been fundamental in recent attempts to reunify lexical semantics (e.g., Jackendoff, 2002). To understand that no real conflict exists, consider a fundamental result of algebraic topology: the simplicial approximation theorem, showing that any continuous function between two Euclidean spaces can be approximated by a mapping between the vertices of some appropriate triangulation of the spaces. As I have already shown, Voronoi tessellations provide a simple model of how categorization subdivides a conceptual space into convex sets. Furthermore, the Voronoi diagram has a dual graph, which is a set of triangles generated by joining contiguous prototypes, that is, the Delaunay triangulation (fig. 16.3). This suggests that prototypes generate a basic triangulation of conceptual spaces, in which they play the role of simplicial vertices. A natural interpretation is that prototypes can provide the corners of the simplicial approximation of a continuous map between mental spaces. The correspondence between prototypes and words (or other lexical elements) then

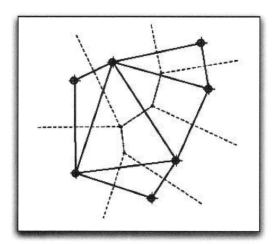

Figure 16.3
A Delaunay triangulation (solid lines) superimposed on a Voronoi diagram (dashed lines).

explains how language can serve as a mediator between conceptual spaces, approximating a continuous function. The result is a great economy in the cognitive resources needed to memorize and process such a function (cf. Edelman's [1999] "chorus function"). For language users, the approximation boils down to remembering and communicating about the prototypes of a Voronoi decomposition of the space. In this way, the approximation can serve as a bridge between the discreteness of language and the continuity of the semantic reaction function. The bottom line is that the mechanisms of language and linguistic categorization are sufficient to approximate continuity with economic discretization. The convexity of spaces plays two important roles here: it ensures triangulability, and it allows reconstructing the behavior of the approximated function as convex combinations of the values of the approximated function in the correspondences of vertices.

It may be necessary to change the grain of the triangulation of the spaces to grant a simplicial approximation. Human categorization systems have different levels of granularity, corresponding to different sets of prototypes (Rosch, 1975, 1978). An approximation between two triangulated spaces may be made finer-grained by further triangulating such spaces. Moving between levels of categorization may ensure that finer triangulations can be constructed and a simplicial approximation achieved. Note that, just as categories can be refined locally, not all simplexes of a simplicial complex need to be further triangulated to achieve the required degree of decomposition. Given a complex, one can leave unchanged one subcomplex while further triangulating another subcomplex.

Notes

1 What Is Semantics?

1. I am not saying the syntax does not contribute to the meaning of a sentence, only that without knowledge of the meaning of the basic words there is no need for syntax.

2. In line with this program, Verhagen (2005, p. 76) writes that "all language, when actually put to use, involves the coordination of one cognitive system with another."

3. In linguistics, the appropriate word for "word" is "part of speech." When I use "word," I normally mean "word stem," since I am not concerned with morphology.

4. They were actually presented as questions.

5. For a presentation of some of the realist semantic theories, see Gärdenfors (2000, chap. 5).

6. Here "communicative expressions" is a summary term for gesture, sign, spoken, and written expressions.

7. This position is clearly expressed by Lewis (1970, p. 19): "I distinguish two topics: first, the description of possible languages or grammars as abstract semantic systems whereby symbols are associated with aspects of the world; and second, the description of the psychological and sociological facts whereby a particular one of these abstract systems is the one used by a person or population. Only confusion comes of mixing these two topics." I disagree. Languages do not exist without their users.

8. This position goes against "modular" theories of language like that of Fodor (1983).

9. Functional linguistics and construction grammar take a middle position. These theories have contributed to the understanding of complex expressions and syntactic structures in close collaboration with the tradition of cognitive linguistics.

10. This criterion is also known as "the mapping problem" (see, e.g., Mintz and Gleitman, 2002, p. 268).

11. Learning the meaning of a word is far from perfect, though, as will be seen in section 2.4. Children often overgeneralize.

12. This is a version of Quine's (1960) gavagai problem.

13. In Gärdenfors (1997, 2000), my fourth criterion was called the communicative question.

14. Communication relies on the mutual assumption that we are talking about the same things; it is too strong to demand that we *know* that we are communicating about the same things.

15. Jackendoff's "cognitive constraint" (1983, p. 16) is closely related, although it focuses on perception.

16. Of course, parallel problems arise for the other senses too: how can we talk about what we taste? How can we talk about the music we hear? Although I focus on vision, the general problem is how meaning can be grounded in perception (Barsalou, 1999).

17. "The world has the structure of language, and language has the forms of the mind" (my translation).

18. Another example is Langacker (2008, p. 30), who says that meaning is conceptualization.

19. Researchers within cognitive linguistics present a multitude of examples of image schemas but hardly analyze which schemas are possible and which are not. Zlatev (1997, pp. 40–44) argues that the notion is used in different ways by different cognitive semanticists.

20. They also claim that most image schemas are closely connected to *kinesthetic* experiences. This is part of the *embodied* aspect of cognitive semantics.

21. In making this analogy, it is important to distinguish between *automatic* attention, where an external event or object attracts your attention, and *voluntary* attention, where you choose to direct your attention to something. The analogy is that language triggers automatic attention shifts in the inner world.

22. Lakoff assumes that the path meaning of "over" is primary. In chapter 11, I argue that it is rather the process (dynamic) meaning that is central. See also Jackendoff (1983, chap. 9).

23. Also see Holmqvist (1993), sec. 1.11.1.

24. Following the convention in linguistics, the asterisk (*) marks a sentence that is judged to be unacceptable.

25. Note, however, that not all metonymies are of visual origin; metonymic relations can also be functional, causal, and so on (Peirsman & Geeraerts, 2006).

26. For an early computational model, see Holmqvist (1993).

27. In Langacker's later work, force, often called energy, figures more prominently.

28. Mandler (2004, viii) argues, however, that infants' minds are more structured than Piaget assumes: "Infants appear to be conceptual beings from the start, without going through an extensive sensorimotor period lacking any conceptual thought."

29. For a presentation of the arguments and counterarguments, see Gärdenfors (2000, sec. 5.7).

30. This can be seen as a special case of the "brain-to-brain coupling" that is highlighted by Hasson et al. (2012).

31. I elaborate on this position in chapter 5.

32. Cf. the *alignment* process studied by Pickering and Garrod (2004).

33. See also Stalnaker (2002) for a different account.

34. The phonology of the word may have changed drastically, though.

2 Conceptual Spaces

1. When Croft writes "conceptual space," he refers to the special case that will be called semantic maps in section 2.9.3.

2. For an extended discussion of the notion of similarity, see Gärdenfors (2000, sec. 4.3).

3. For a discussion of the role of dimensions in scientific theories, see Gärdenfors and Zenker (2013).

4. Melara (1992, p. 274) presents the distinction as follows: "What is the difference psychologically, then, between interacting [integral] and separable dimensions? In my view, these dimensions differ in their similarity relations. Specifically, interacting and separable dimensions differ in their degree of *cross-dimensional similarity*, a construct defined as the phenomenal similarity of one dimension of experience with another. I propose that interacting dimensions are higher in cross-dimensional similarity than separable dimensions."

5. Yes, I have an instrumentalist view on the role of scientific theories.

6. Huth, Nishimoto, Wu, and Gallant (2012) provide some evidence that a semantic space is organized into continuous dimensions that cover much of the visual and nonvisual cortex.

7. For a discussion of different origins of quality dimensions, see Gärdenfors (2000, sec. 1.9).

8. Cf. Waismann (1968, p. 124).

9. See Gärdenfors (2000, sec. 3.4) for a discussion of convexity and related concepts.

10. Convexity can be defined for graphs and orderings, among other things.

11. Object categories correspond to what in philosophy are sometimes called *sortals*.

12. Cf. Clausner and Croft (1999, p. 7). Also see the example of *apple* in Gärdenfors (2000, sec. 4.2.1) and Langacker's (2008, p. 47) list of domains for *glass* (this list contains examples that are not counted as domains in this book).

13. The role of concepts in reasoning is elaborated in Gärdenfors (2000, chap. 4).

14. The squash value (denoted by s) of an object is measured by the formula $s = (s_1 - s_2) / s_2$, where s_1 is a height before squash and s_2 is a height after squash.

15. For example, in a Euclidean space (and other metric spaces), one can calculate the center of gravity.

16. In chapter 13 (and in Gärdenfors, 2000) I argue that the region assigned to a linguistic expression may not be constant but in general depends on the context.

17. A related theory, called *packing theory*, is presented by Hidaka and Smith (2011). Their theory demands more of memory, though, to represent categories.

18. This constraint generates a *lattice* of concepts.

19. More precisely, they should understand that volume = height × width × width. The age at which children manage the conservation task has been debated, and it seems to depend on how the task is set up. Piaget claimed that children manage the task around seven years of age, but in many contexts younger children demonstrate awareness of the conservation of volume.

20. This and the following section are based on material from Gärdenfors and Löhndorf (forthcoming).

21. In older linguistics, the notion of a *semantic field* has been popular. Brinton (2000, p. 112) defines it as follows: "A semantic field denotes a segment of reality symbolized by a set of related words. The words in a semantic field share a common semantic property." I use the notion of a domain in a narrower sense.

22. This allows them to claim that an "image schema is a subtype of domain" (Clausner & Croft, 1999, p. 4).

23. I am grateful to William Croft for pointing this out.

24. Evans and Green (2006, p. 234) list space, color, pitch, temperature, pressure (force), pain, odor, time, and emotion as basic domains. Except for pain, they are all among the basic domains treated in this book.

25. This distinction has been copied by several other researchers (Clausner & Croft, 1999, pp. 7–13; Evans & Green, 2006, p. 236; Croft & Cruse, 2004, p. 21).

26. Langacker (personal communication) says that he does not intend the distinction between locational and configurational domains to be exclusive.

27. Nosofsky (1992) provides a survey of mathematical models in this area.

28. Langacker's position is repeated by Evans and Green (2006, p. 236).

29. Langacker (1987, p. 151) presents these two dimensions, too, but he does not seem to regard them as exhaustive.

30. This section builds on Gärdenfors (2000, sec. 3.10).

31. The rectangles have the same *shape* if $|ab| / |ac| = |ef| / |eg|$, that is, if the quotients of the lengths of the sides are the same.

32. On E, one can define a metric d, by $d(<a,b,c,d>, <e,f,g,h>) = \sqrt{((|ab| - |ef|)^2 + (|ac| - |eg|)^2)}$.

33. I discuss the role of invariances further in chapter 11.

34. This process is called "relation linking" by Wisniewski (1996).

35. The ideas in this section were developed as part of the VAAG project. I thank the project members, as well as Igor Douven and Paul Egré for helpful comments.

36. The literature contains models that build on fuzzy representations of membership rather than on probabilities. The points I make concerning vagueness are independent of which way one chooses to mathematically represent membership.

37. Feature analysis stems from phonology, where phonemes are characterized by a finite number of binary features.

38. We can also see the *scripts* of Schank and Abelson (1977) as a variation of frames.

39. Croft (2001, p. 92) calls semantic maps "conceptual spaces," which is confusing in the present context. Hence I stick to the term "semantic map."

40. This hypothesis can be strengthened into the hypothesis that all semantic maps are *convex*. The difference can be made in Haspelmath's map: 3–4–6–8 is connected, but not convex, since 5 is between 3 and 8. The hypothesis is not fully valid, since Haspelmath's investigation contains a language with 4-5-8 and one with 4-6-8. However, all areas are "star-shaped" in the sense that a subset C of a conceptual space S is said to be *star-shaped with respect to point p* if, for all points x in C, all

points between x and p are also in C. See Gärdenfors (2000, sec. 3.4) for a more detailed presentation of the meaning of connected, star-shaped, and convex regions.

41. Cf. Sanso's (2010) distinction between first- and second-generation semantic maps.

42. Regier, Khetarpal, and Majid (forthcoming) present an interesting algorithm for inferring a graph-based map from linguistic data (inspired by the problem of inferring a social network from disease outbreaks).

43. Recall that Croft uses "conceptual space" instead of "semantic map."

3   The Development of Semantic Domains

1. The arithmetic domain is important in development, but I do not consider the semantics of numbers in this book, since they are not central to my lexical analysis.

2. Although they write about conceptual categories instead of domains, Smiley and Huttenlocher (1985, p. 25) hold a similar position: "Although input helps children acquire particular words, the order of acquisition and the nature of word meanings chiefly reflect the accessibility of certain conceptual categories at different points in time."

3. Attributing intentions presumes a goal domain, which is treated in section 3.2.6.

4. Similarly, Clark (1996, pp. 33–35) writes about coordination of goals.

5. Southgate, Senju, and Csibra (2007, p. 588) argue that the "where" question that is involved in most linguistic versions of the false-belief task "is prematurely interpreted by young children as referring to the location of the hidden object, rather than the location of the actor's subsequent actions."

6. Of course, a representation of more nuanced emotions may involve further dimensions.

7. Many of the principles of rhetorics concern how the speaker can create shared emotions with the audience rather than shared information.

8. However, that warning calls are used for deception indicates that the referential function is presumed by the deceivers.

9. Note that the emotions associated with these four patterns correspond roughly to the four quadrants of figure 3.1.

10. Jackendoff (1987a) calls this "spatial representation."

11. A special case is *altercentric* space, that is, the ability to see things from the point of view of somebody else. The terms "left" and "right" may be ambiguous, depending on whether one chooses an egocentric or an altercentric perspective.

12. Haviland (2000) provides several examples of languages that use "cardinal" direction in the sense that they always refer to the allocentric space using "north," "south," and so on. In these languages, storytellers "assiduously orient pointing gestures in the 'correct' compass directions" (p. 13).

13. In contrast to this position, Mandler (2004, p. 101) argues that "it is not necessary to build in innate knowledge about force and related mechanical properties." However, she also says that "it may be that a notion of causal force is not primary but instead is derived from analysis of the transfer of motion from one object to another, in conjunction with bodily experience of pushing against resistance and being pushed" (p. 101). Although a force domain may not be innate, Mandler's position includes bodily experiences of forces and does not contradict the assertion that a general force domain develops early.

14. This distinction mirrors the products of the dorsal and ventral streams of visual processing in the brain. Landau and Jackendoff (1993) speculate about how the two systems are reflected in the expressive capacities of lexicon for object names and for spatial relations.

15. Jackendoff (2007, chap. 9) also emphasizes the importance of the value domain for conceptualization.

16. These types of values (and some others) are also discussed in Jackendoff (2007, chap. 9).

17. Wolff (2007, 2008) argues, on the basis of his experimental evidence, that subjects treat intentions as causes that are analogous to forces.

18. As I show in chapter 10, this model of intentions is the same as the model of events—with the exception that the action involved in an intention is only planned. In other words, an intention is an imagined event.

19. Polunin (2009) argues that in many situations, several temporal processes are represented simultaneously.

20. In fact, most of the time, usage shows a slow decline, in part because as more words are learned, the proportion for any particular word decreases.

21. ChildFreq was developed by Bååth (2010). It is available at http://childfreq .sumsar.net.

22. It is difficult to identify any words that indicate the understanding of emotions (empathy), since this capacity develops well before words are learned, and there is much nonverbal communication about emotions.

23. Cf. Nelson (2005) on how children learn the word "know."

24. I present further data concerning dimensional adjectives in chapter 7.

25. I expand this analysis in chapter 13. See also Gärdenfors (2000, p. 176).

26. "When people cannot change things, they change the words" (my translation).

27. Clark (1996, p. 56) calls the Chomskian perspective the *production tradition* (focusing on the products of language) and the perspective that puts pragmatics in focus the *action tradition*.

28. There are, of course, other ways of sharing the physical domain, for example, by using auditory, tactile, or olfactory stimuli.

4   Pointing as Meeting of Minds

1. To a large extent, this chapter is based on Gärdenfors and Warglien (2013).

2. This proposal also fits well with gesture theories of language origin such as that of Corballis (2002).

3. Pointing is not done with the finger in all cultures; sometimes it is done with the chin.

4. The first corresponds to what Tomasello et al. (2007) call the expressive and the second and third to what they call the informative subtypes.

5. In line with this, Tomasello (2001, p. 134) says that the "social-pragmatic approach" "begins by rejecting truth-conditional semantics in the form of the mapping metaphor (the child maps word onto world), adopting instead an experientialist and conceptualist view of language in which linguistic symbols are used by human beings to invite others to experience situations in particular ways."

6. I discuss the role of projections further in chapter 12.

7. This occurs frequently in sign language, where characters in a discourse are placed in the space in front of the signer. These places are used like definite pronouns, and the signer may point to them for reference.

8. I am grateful to Ingvar Johansson for bringing out this parallel.

9. See, e.g., the table in Clark (1996, p. 188), for different ways of describing-as, indicating, and demonstrating.

5   Meetings of Minds as Fixpoints

1. Parts of this chapter, in particular sections 5.4–5.8, build on Warglien and Gärdenfors (2013).

2. As I will show, the notion of fixpoints in semantics is related to the equilibriums in language games. Mathematically, a fixpoint for a function f is a value x such that $f(x) = x$.

3. An even earlier formulation can be found in Dewey (1929, chap. 5).

4. Mead (1934, p. 76) adds that meaning is "a development of something objectively there as a relation between certain phases of the social act; it is not a psychical addition to that act and not an 'idea' as traditionally conceived."

5. Hasson et al. (2012, p. 120) show that a coupling of brain activities can be detected in two participants who communicate. They conclude that coupled brains "can create new phenomena, including verbal and nonverbal communication systems and interpersonal social institutions, that could not have emerged in species that lack brain-to-brain coupling."

6. This form corresponds to the language games introduced by Wittgenstein (1953).

7. Cf. Bara's (2010, pp. 195–196) three types of failures.

8. Cf. Krifka (2012b, p. 3).

9. Lawler (1973, p. 112) notes that "indefinite generics seem most natural in definitional sentences, or ones used somehow to identify the nature of the thing specified by the generic by means of properties peculiar to it; they are less acceptable when an accidental quality is predicated on them."

10. Note that pointing, unlike language, is a continuous way of referring to the outer world: the direction of the finger is continuously variable.

11. "Learning to speak is learning to translate" (my translation).

12. In some cases, children also undergeneralize by only using a word in limited context (this phenomenon is more difficult to note for the listener, though). Vervet youngsters also overgeneralize to other similar phenomena and call out too often. That the adults ignore a warning call for eagle from a youngster may have the result that when the kid sees a serpent eagle next time, it has learned not to emit the warning call for serpent eagles (which is a species that is not dangerous for vervets). Cheney and Seyfarth (1990) have studied how vervets learn to narrow their use of the different warning calls.

13. As I show in section 5.5, the convexity of their conceptual spaces may also help in reaching a solution.

14. "Speech comes half from the one who speaks, half from the one who listens" (my translation).

15. It is only assumed that the similarity is a decreasing function of the distance in the color space between nature's choice and r's choice.

16. That a space S is compact means that every infinite subset of S has a limit point in S. In brief, the space contains its borders.

17. Jäger (2007, pp. 562–563) writes that the result can be interpreted in two ways: "Under one interpretation, the convexity of cognitive categories is not so much a

property of cognition but rather a consequence of a positive feedback loop in communication. . . . The notion of a signaling game can also be interpreted in a purely cognitive way though. Under this interpretation, 'meanings' in the sense of the model are perceptual stimuli and 'signals' are higher-lever representations of categories. A move of a 'sender' would be the categorization of some perceptual stimulus, and a move of a receiver the invocation of a prototypical exemplar of a given category. . . . Of course these two interpretations, the communicative and the cognitive one, are not mutually exclusive and may both be partially true."

18. This use of fixpoints resembles the attempt made by game theorists to define equilibrium as a state of mutual compatibility among individual strategies (see Parikh's [2010] equilibrium semantics for a related approach).

19. Parikh (2010) shows more complex examples of how payoff dominance may select meaning.

20. I present an account of metaphorical composition in section 13.3.

21. Chierchia (1995, pp. xii—xiii) presents two approaches to the updating process: "On the view of meaning as content, we might simply say that the utterance of a declarative sentence in a given context will naturally prompt us to enter its content into the common ground—and that is the end of the story. The second way to go is to view sentences as actually having as semantic values functions from information states to information states. On this second view, context updating would be an integral part of the compositional system of meanings." I introduced the second process in a more formal setting in Gärdenfors (1985).

22. A related theory is that of *mental spaces* proposed by Fauconnier (1984). Despite its name, it does not contain any geometric structures.

23. For a mathematical treatment of this, see Warglien and Gärdenfors (2013).

24. This accords with Clark's (1996, p. 24) proposition 6: "The study of language use is both a cognitive and a social science."

25. In one of his three tenets, Clark (1992, p. xv) formulates this idea as follows: "In language use, speaker's meaning is primary, and word or sentence meaning are derivative."

6   Object Categories and the Semantics of Nouns

1. There are other opinions: Gil's (1994) work on Riau Indonesian suggests that it has no distinctions between different kinds of words. In chapter 12, I discuss the universality of word classes in more detail.

2. There exist, however, other identifiable word classes, for example, numerals and articles.

3. Langacker (2008, chap. 4) pursues a similar program.

4. As I argue in section 6.7, so-called relational categories are of a different nature.

5. The meronomic aspects of categories were not considered in Gärdenfors (2000), and they represent an extension of the notion of an object category, compared to what I have presented there.

6. In Jackendoff (1991, sec. 5.3) this shows up as the COMP function.

7. In fact, a large part of formal education, in particular in the sciences, is devoted to introducing new theoretical domains that can explain phenomena in perceptual domains.

8. Smith and Heise (1992, p. 242) confirm the role of prominence: "Perceived similarity is the result of psychological processes and it is highly variable. The central process that changes perceptual similarity is attention. The perceived similarity of two objects changes with changes in selective attention to specific perceptual properties." Compare Nosofsky (1986). I discuss this topic in greater detail in Gärdenfors (2000, sec. 4.2).

9. The domains with the strongest prominence can be considered to represent the essential properties of the category. By "essential" I mean only that the domain is part of the core meaning of a concept. I do not subscribe to any form of ontological essentialism. See Gärdenfors (2000, chap. 3) for a discussion of essentialism.

10. Croft and Cruse (2004, p. 18) discuss a similar example concerning the difference between "land" and "ground."

11. Pustejovsky (1995) introduces the type-theoretical construction of "dot objects" to handle such ambiguities. In my analysis, the type sorting emerges from an analysis of the prominence of domains, and hence no additional construction is needed.

12. I discuss this topic in greater detail in Gärdenfors (2000, sec. 4.6).

13. Jackendoff (1991) introduces the partitive function as one among many semantic functions, but he does not discuss how parthood is to be determined.

14. See also Jackendoff (2012, p. 1141).

15. Example (6.3) can make sense if "went through it" means "went through the window by crushing the pane."

16. This corresponds to the notion of *domain ranking* in Langacker (1987, p. 165). Langacker also writes about *salience* instead of prominence.

17. Object categories that are determined by their *functional* properties will be treated in section 8.5.

18. The correlations show up as our *expectations* about the world. Note that the inferences we draw on the basis of our expectations do not depend on linguistic (symbolic) representations.

19. This kind of concept combination will be treated in chapter 13.

20. In Gärdenfors (2000, secs. 6.3–6.4), I write about this kind of "analytic truth" that is dependent on the choice of a particular conceptual space.

21. All "is-a" relations of semantic networks emerge in this way.

22. What counts as the basic level depends partly on your culture and partly on your area of expertise. For example, a botanist would hardly put "flower" on the basic level of object categories.

23. I discuss the role of meaning postulates more extensively in Gärdenfors (2000, sec. 6.3), in particular in relation to Carnap's philosophy.

24. In a very similar vein, Langacker (1991, sec. 2.3.2) treats the instantiation of a type as just a special case of the relation between types and subtypes. It is interesting to note that a similar construction occurs in Montague's (1970) type construction of individuals, where an individual is defined as a maximal set of compatible properties (see also Lewis 1970, p. 53).

25. Mead (1934, p. 78) writes: "Language does not simply symbolize a situation or object which is already there in advance; it makes possible the existence or appearance of that situation or object, for it is part of the mechanism whereby that situation or object is created."

26. Leibniz's law concerning the identity of indiscernibles obtains a special meaning in the present context: if two objects are assigned the same values in all relevant dimensions, the objects are construed as identical. In other words, that two objects are discernible means that there is some dimension along which they differ. For physical objects, this may be just the location in space. Also cf. Parsons's (1980) principle that no two objects have exactly the same nuclear properties (where his meaning of nuclear property comes close to the notion of property in sec. 2.2).

27. Thus I have no need for possible worlds or any other form of modal realism to account for possible objects.

28. Cf. the analysis of intentional identity in section 5.8.

29. A technical comment is that I am not assuming that the time dimension is necessarily represented for a physical object, only that there is some way of representing change. The reason for this will become clear in chapter 8. Continuity can still be defined as continuity of change.

30. Since we have only partial information about the trajectories, there is no paradox in the fact that two objects are identical, but we do not know that they are identical. Looking at how objects are represented cognitively thus resolves the morning star paradox.

31. I call them characteristics because they are not properties in the technical sense of the word used in this book. I call them cognitive because they are examples of

how the mind structures the world (cf. Kant's "das Ding für uns"). I do not claim that the "real" world ("das Ding an sich") has the same characteristics.

32. In the case of mass nouns, one may find marginal exceptions, for example, the milk in the coffee is normally thought of as being at the same place as the coffee.

33. Lyons (1977, p. 446) says that this classification is not intended to be exhaustive. Numbers, sets, and so on, fall outside.

34. *Substances* such as sand, water, and air may not have a well-circumscribed location and are therefore different from objects. I return to substances in relation to mass nouns in section 6.8.

35. But "summer" also has uses that require (spatio)temporal continuity: "We will have a better vacation next summer."

36. A categorization of the same kind comes from Pustejovsky (2001). He separates the domain of individuals into three distinct levels: (i) *natural* types (natural-kind concepts referring only to his formal and constitutive qualia roles); (ii) *artifactual* types (concepts referring to purpose or function; and (iii) *complex* types (concepts referring to an inherent relation between types).

37. I discuss these so-called *thematic roles* further in chapter 9.

38. Yet "robber" has the properties of an *agent*. These properties will be discussed in chapter 9.

39. Langacker (1991a, p. 63; see also 1991b, vol. 2, pp. 15–16) defines nouns in terms of a *single* domain—and his examples are selected correspondingly. But on p. 74 he defines "quality space" as a domain. His quality space is closely related to the object category space presented in this book.

40. Cf. Zhu and Yuille's (1996) model of parts presented earlier in this chapter. See also Langacker (1991a, pp. 69–70; 2008).

7 Properties and the Semantics of Adjectives

1. Linguists have debated whether there are languages that do not have adjectives. For example, some researchers have claimed that in Mandarin all adjectives are verbs. Dixon (2004, p. 1) reviews the evidence and concludes that something that has the role of modifiers can be found in all languages: "In some languages, adjectives have similar grammatical properties to nouns, in some to verbs, in some to both nouns and verbs, and in some to neither. I suggest that there are always some grammatical criteria—sometimes rather subtle—for distinguishing the adjective class from other classes."

2. Note that, in contrast to element identification, an indefinite determiner is used for kind identification.

3. Dietz (2013) formulates a version of the Voronoi tessellations that is based on a comparative concept. It is more general than the one I presented in chapter 2, and it provides a framework for graded categorization that has my version as a limiting case.

4. In fact, Jäger used the CIELab space instead of the double cone.

5. A technical detail is that Jäger (2010) uses linear separators for the color space. This corresponds to applying a Euclidean metric in defining the partitioning of the space. In Gärdenfors (2000, sec. 3.5), I argued that for the color space, it would be more natural to use *polar* coordinates that generate a different metric and thus a different partitioning. It would be interesting to investigate whether this metric would give an even better fit to the data.

6. As an illustrative example that these are the most central domains, Dixon (1977) showed that Igbo has only two adjectives for each of the classes dimension, age, value, and color and no more. Croft (2001, p. 99) writes that they are the most prototypical adjectives.

7. Thanks to Wilhelm Geuder for this example.

8. See also Smith and Sera (1992, p. 132). Mintz and Gleitman (2002, p. 269) present a similar proposal: "Glaring asymmetries in noun vs. adjective (and verb) frequencies in novice vocabularies . . . persist until about their third birthday. . . . One potential explanation for why acquiring adjectives is hard has to do with the possibility that they fall into a variety of conceptual classes whose conflation under a lexical categorization . . . is more arbitrary than natural." Their phrase "conceptual class" corresponds to my "domain."

9. For a definition of the establishment period, see section 3.3. The pairs that are included in the list are based on Deese (1964), but some additional pairs have been added. It includes the pairs where the frequency of use is sufficiently high for both words and where the development curve allows a reasonable identification of an establishment period. Some antonyms that also have frequent uses as nonadjectives, for example, "top–bottom," are also excluded. Admittedly, this selection introduces a certain bias.

10. The positive member, for example, "hot," "tall," "high," "heavy," of each pair is the term that is used in a neutral question (how hot is the weather?). Furthermore, the nominalization of a dimension is derived from the positive form (what is the height of the house?). Some pairs have no clear positive member.

11. Clark (1972, p. 755) also considers the order of acquisition of antonym generation in a set of pairs of adjectives. There is a strong correlation between her results and the results presented in table 7.1.

12. Deese (1964, p. 355) warns: "The associative independence of the contrasting pairs in the present data questions the advisability of trying to reduce the 20 scales

of the semantic differential to a smaller number (usually three) of orthogonal factors."

8   Actions

1. To be accurate, van Gelder and affiliates avoid using the notion of representation, since they associate it with the symbolic approach to cognition.

2. But see Zwarts (2005) on the role of forces in the meaning of prepositions. Prepositions will be treated in chapter 11.

3. I also use scalar multiplication of vectors in my analysis of the meaning of adverbs in section 12.3.

4. Jackendoff (1990, chap. 7) presents a slightly more detailed analysis based on causation rather than forces. A remark in passing is that he distinguishes between causes as application of force with a successful outcome and causes as application of forces with undetermined outcome (p. 132). However, the notion of "successful outcome" seems to bring in an element of intentionality that I do not think is warranted in all cases of force dynamics (see sec. 10.6).

5. More precisely, Marr and Vaina (1982) only use differential inequalities, expressing, for example, that the derivative of the position of the upper part of the right leg is positive—in the forward direction—during a particular phase of the walking cycle.

6. The hypothesis about actions is also supported by some neurophysiological data, for example, Flanders, Tillery, and Soechting (1992); Bullock, Cisek, and Grossberg (1998).

7. For more examples from phonetics, see Gärdenfors (2007b).

8. Of course, if several forces are involved—as in bodily motion—the relevant action space will be multidimensional, and identifying the convex regions may be more complicated.

9. Further support for prototypicality in movement patterns relative to gender, emotions, and so on can be found in Troje (2008).

10. Giese et al. (2008) go further in analyzing the metric space that underlies the perception of body movements. Like Giese and Lappe (2002), they use morphs of prototypical movement patterns. Starting from videos of persons who are walking, marching, or running, they created moving-dot films and then linear morphs of these three actions. Subjects were shown two pairs of films from this set and asked to judge which pair of movements is the most similar. To compare with physical distance, Giese et al. developed a measure for determining the distance between any two of the actions. They then compared these "physical" measures to the results of

the generated two-dimensional perceptual space and found an extremely close match between the relative locations of the different movements in the two spaces. Giese et al. (2008, p. 12) conclude that "perceptual representations of complex body movements are *veridical* in that they closely reflect the metric of the movement in the physical world."

11. The relative weights of the dimensions may be an artifact of the selection of actions included in the experiment. If dimension 2 is given lower weight, the regions for run and jump/hop will become "rounder."

12. One example I will return to is Wolff (2007, 2008), who, in a series of experiments, studied the perception of causation on the basis of the role of forces.

13. For example, there are chairs in the shape of a large hand directed upward.

9 Events

1. This chapter builds to a large extent on Gärdenfors and Warglien (2012).

2. I return to the notion of a construal in section 9.6.

3. Jackendoff's formalism (e.g., 1983, 1991) goes beyond the purely symbolic, since he also represents spatial structures.

4. The class of events has a prototype structure. Therefore I do not present a definition of events based on necessary and sufficient conditions.

5. Later I will show that there are events such as *remaining* in a state because the force vector is balanced by a counterforce that leads to no change.

6. A difference between the model I present here and that of Croft (2012a) is that Croft allows events to form causal chains involving several subevents. This is illustrated by his analysis of "Sue broke the coconut for Greg with a hammer" (see the figure on p. 214 in Croft, 2012a). Here Sue applying force is a subevent that causes the subevent of the hammer impacting the coconut that in turn causes the subevent of the coconut breaking, which finally causes the subevent of Greg benefiting from the broken coconut. In contrast, I analyze this as a single event containing several roles.

7. Da Silva Sinha et al. (2012) argue that the Amazonian language Amondawa can only express temporal relations in terms of event, but the language does not have any way of referring to time as such, that is, as a separable time dimension.

8. It has actually been suggested that time intervals can be construed from the order of events (Reichenbach, 1928; Thomason, 1989). Of course, the time dimension is necessary for the analysis of tense. That is a topic, however, that I do not treat in this book.

9. However, I do not exclude the possibility that the time dimension may be part of an event representation. Zwarts (2008) notes that for some verbs, such as "postpone" and "antedate," the time dimension is required.

10. There are also events involving *joint actions*. Consider two people wrestling, dancing, or arguing. Forces alone cannot determine who is the agent. One agent's force is another's counterforce. I will not try to analyze joint actions here.

11. The domain-neutral notion of *power* may have cognitive priority over force (Jammer, 1957).

12. Note that the two-vector model of events—including agents—satisfies most of Dowty's (1991, p. 572) criteria for a proto-agent and a proto-patient. The proto-patient undergoes a change of state, can assume the role of incremental theme, and is causally affected by another participant. The proto-agent causes an event or change of state in another participant. I return to the criterion that the proto-agent is sentient in my analysis of perception verbs.

13. I introduced the goal domain in section 3.2.6.

14. See Levin and Rappaport Hovav (2005) for a survey. The analysis presented here generates a partial hierarchy of thematic roles. Furthermore, the model is able to cover the "subeventual analysis" of Levin and Rappaport Hovav (2005, p. 112).

15. By "event structure," they refer to the ACT-CAUSE-BECOME analysis presented earlier. Similarly, Jackendoff (1991, p. 13) writes: "Thematic roles are treated as structural positions in conceptual structure, not as an independent system of diacritics (or case-markers)."

16. The category becomes more complicated to describe if force patterns and paths and not just vectors are involved in the changes.

17. Later Jackendoff has further extended his analysis of verbs, for example, to mental and social verbs in Jackendoff (2007), in a way that goes beyond the localist hypothesis.

18. Again, a more general description would involve paths and not just vectors.

19. Langacker's (1987, sec. 7.2) analysis of aspect contains basically the same distinctions as Croft's (2012).

20. Langacker's (1986b) model has precedence, but it is less developed. He has no notion of vectors, and he writes about energy instead of forces.

21. Some support for this position comes from Sinha et al. (2011) and da Silva Sinha et al. (2012), who argue that the Amazonian language Amondawa does not have an explicit time domain in the language but only expresses events. Furthermore, the standard metaphor mapping from space to time is not a feature of this language.

22. An analogy to a construal is perhaps visual perception leading to judgments of the form "category X is at location Y." Searle (1983, p. 40) writes: "The content of the visual perception . . . is always equivalent to a whole proposition. Visual experience is never simply *of* an object but rather it must always be *that* such and such is the case." He says that his experience of a station wagon must also be an experience of, for example, a station wagon in front of me. Given the thesis about construals, however, the analogy does not capture all kinds of event construals, in particular not the dynamic aspects.

23. Talmy (1988, p. 61) summarizes the position succinctly: "All of the inter-related factors in any force-dynamic pattern are necessarily copresent wherever that pattern is involved. But a sentence expressing that pattern can pick out different subsets of the factors for explicit reference—leaving the remainder unmentioned—and to these factors it can assign different syntactic roles within alternative constructions."

24. Croft (2012a, p. 256) describes the passive voice as a *deprofiling* of the causal chain from the agent to the patient. This can be expressed in my terminology by saying that the patient is made the focus (or topic) of the event.

25. Jackendoff (1987c, p. 380) also accords: "*Subject* is a syntactic relation, not a conceptual one, and syntactic subjects can hold a variety of θ-roles."

10   The Semantics of Verbs

1. Cf. Perlmutter and Postal's (1984, p. 87) Universal Alignment Hypothesis.

2. Also within philosophy there is recent criticism of the general use of predication. See, e.g., Ben-Yami (2004) and McKay (2006).

3. It cannot be the only function, however, because some verbs do not admit imperatives (Culicover & Jackendoff, 2005, sec. 12.3).

4. A fascinating proposal is that of Thom (1970, p. 232), who claims that any basic phrase expressing an interactive process can be described as one out of sixteen fundamental types. Among these types one finds *begin*, *unite*, *capture*, and *cut*. The sixteen types are derived from some deep mathematical results concerning morphologies within catastrophe theory. Now, even if there are excellent mathematical reasons why there exist exactly sixteen types of interaction, it is not obvious that they correspond neatly to cognitive representations, though such a correspondence would be theoretically very gratifying.

5. Possible exceptions to this general rule—which I discuss later—are verbs that describe changes in ontology (see sec. 10.4) and verbs like "give" that describe intentional actions involving recipients (see sec. 10.6).

6. A similar point is made by Kaufmann (1995, p. 85).

7. Jackendoff's (1983, p. 179) distinction between "what happened was" (for events) and "what x did" (for actions) is related, except that his event description involves both force and result elements.

8. However, as Levin and Rappaport Hovav (forthcoming, pp. 11–12) note, uses of "climbing down" exist for trains, buses, and planes. Krifka (2012a) notes interestingly that there has been a historical change of the meaning of "climb." Combining "climb" with downward movement is a relatively recent phenomenon. He writes that "the frequency of this collocation in English books rose sharply, more than tenfold, from <0.02 per million before 1850 to >0.2 per million after 1910, whereas the frequency of 'climb up' rose only slightly from about 0.6 per million to about 1 per million in the same time. This is consistent with a view that "climb" used to specify upward movement before 1850, a meaning component that dropped out after 1900."

9. What Kiparsky (1997, p. 17) calls disjunctive meaning is thereby not disjunctive at all: instead the uses have a prototype structure with a central meaning. My use of "central" meaning is clarified in chapter 11.

10. This is clear from the lexicon definition cited by Kracht and Klein (2012).

11. Even though Kaufmann (1995) uses a different terminology, I believe that most of the constraints she formulates for which verbs exist can be derived from the thesis about verbs and the single-domain thesis.

12. I return to the role of intentionality in creating double meanings in section 10.6.

13. Rappaport Hovav and Levin (2010) distinguish between what a verb lexicalizes and what can be inferred from a particular use of that verb in a context. According to them, the criterion for lexicalized meaning is constancy of entailments across all uses. This still excludes the possibility that a verb is lexicalized as both manner and result. My version preserves the idea that in any instance of use of a verb, it is either manner or result, but I do not exclude that both meanings are lexicalized.

14. Following the Gricean terminology, they are also called conventionalized implicatures.

15. As an example, Traugott and Dasher (2002, pp. 36–37) show how the expression "as long as" changes from a purely spatial meaning, to a phase where a spatial meaning coexists with a temporal one, and finally to the present-day situation where only the temporal meaning is active and the spatial one is forgotten.

16. Rappaport Hovav and Levin (2010, p. 21) claim that "manner verbs are found with unspecified and non-subcategorized objects in non-modal, non-habitual sentences, result verbs are not." Thus, for example, "*The toddler broke." Kaufmann (1995, p. 70), presents four criteria for German verbs.

17. I analyze the meaning of "against" in the following chapter.

18. In Rappaport and Hovav's (1998) analysis that was mentioned in section 9.1, the verb "break" is represented as "become broken," which also brings out a connection between result verbs and adjectives.

19. Notice that this construction is different from "the hitting hammer," "the pulling elephant," and so on, where manner verbs can be turned into adjectives denoting an activity that becomes a property.

20. Goldberg (2010) makes a similar point concerning "fry."

21. Many of these higher-level changes could be expressed as state transitions. Such transitions may be explained by the discontinuous effects on the patient space of continuous changes in a force parameter.

22. The copula "is" is a generic stative verb that goes together with an adjective or some other way of describing a property.

23. There are other options: in Russian, natural nonintentional forces are normally expressed as obliques with an instrumental case marking (Croft, 2012a, p. 264).

24. This is reminiscent of Pustejovsky's dot-events, although the model I develop does not require any type-theoretical constructions.

25. Rappaport Hovav and Levin (2010, p. 37) claim that "cut" is a result verb. In the nonintentional sense, it is indeed a manner verb, but in its intentional sense, it is a result verb. The same holds for many verbs with double readings.

26. Which verbs can be used unaccusatively varies between languages. For example, there is no unaccusative use of "bake" in Swedish.

27. Garey's (1957, p. 106) original definition of a *telic* verb is "expressing an action tending toward a goal—envisaged as realized in a perfective tense, but as contingent in an imperfective sense." Here "envisaged" may just as well be replaced with "construed." Later, via the influence of Vendler (1957), "telic" came to mean "with an endpoint," but having an endpoint is not necessary for an accomplishment (see, e.g., Goldberg's [2010] examples 8a and 8b). My use of "intentional verb" thus brings back Garey's meaning of telic.

28. Thus I cannot fully agree with Goldberg's (2010) analysis, according to which the two subevents can be identified because "the resulting state does not completely overlap temporally with the activity." Rather, the crucial factor is that the resulting state is intentional.

29. Note that if "imprison" is replaced with the superordinate verb "house," (34) becomes semantically acceptable, because a prison is canonically used to house people.

30. In their experiments, Huttenlocher et al. (1983) distinguish between change and nonchange verbs, but the distinction closely maps on that between result and manner verbs, except that stative verbs such as "stay" are classified among the nonchange verbs.

31. For a detailed analysis of basic verbs of possession, see Viberg (2010).

32. Thus "I hear" and "I see" are decausativizations just like "the window breaks." The difference is that the patient now is a sentient being.

33. The distinction was first analyzed by Gruber (1967), albeit in a grammatical setting.

34. In line with my analysis, Croft (2012a, chap. 6) describes the perceptual verbs as involving a bidirectional transmission of force in mental events.

35. No such polysemy exists in Italian: *sentire un odore* (to feel an odor) is distinguished from *odorare* or *annusare* (to smell actively).

11   The Geometry of Prepositional Meaning

1. I wish to thank Joost Zwarts for helpful discussions on this chapter and for directing me to central references. Much of the material in sections 11.3–11.6 is also included in Gärdenfors and Zwarts (submitted).

2. I hinted at this analysis already in Gärdenfors (2000, sec. 5.3).

3. Van der Gucht et al. (2007) derive these positions historically from Locke ([1690] 1959) and Leibniz (1765) respectively.

4. They argue that Lakoff (1987) is too lax in allowing new meanings in his analysis of "over."

5. When the poet and Nobel laureate Tomas Tranströmer writes about a house in a storm that it "feels its own constellation of nails holding the walls together," the first metaphor of the house feeling is fairly standard, but seeing the nails of the house as a constellation of stars involves a cognitively much less accessible mapping (which, of course, is part of its beauty).

6. I have already discussed this criterion in connection with semantic maps in section 2.7.3. It has the consequence that all meaning structures will be *star shaped*, with the prototypical meaning in the center (for a definition, see Gärdenfors, 2000, sec. 3.4). For example, all the semantic maps in Haspelmath's (1997) investigation of indefinite reference are star shaped.

7. There are also "epistemic" prepositions such as "despite," "except," and "regarding."

8. In support of the idea that our brains exploit polar coordinates, Gallant et al. (1993) show that more neurons in area V4 in macaque monkeys are sensitive to polar gratings than to Cartesian gratings.

9. See Gärdenfors (2000, sec. 3.5) for further details.

10. Coventry et al. (1994) suggest that "beside" involves a *functional* component, because subjects judge "beside" as more appropriate when trajector and landmark are functionally related, for example, a jug and a glass. If this is the case, then a spatial analysis is not sufficient.

11. The strict definition would enforce the requirement that goal and source prepositions are *transitions* between two phases and that route prepositions have exactly one part in the middle of the [0,1] interval to which the locative condition is applied, while the initial and final phase do not satisfy that condition.

12. This definition can be generalized by allowing k to vary continuously as i varies between 0 and 1.

13. Proof: Suppose that p_1 and q_1 from P and Q respectively are conjoined to $p_1 \cdot q_1$ and that p_2 and q_2 from P and Q are conjoined to $p_2 \cdot q_2$. For any k, $0 \leq k \leq 1$, define p_3 by $p_3(i) = kp_1(i) + (1 - k)p_2(i)$ for all i $0 \leq i \leq 1$ and q_3 by $q_3(i) = kq_1(i) + (1 - k)q_2(i)$ for all i $0 \leq i \leq 1$. P and Q are convex, so p_3 is in P *and* q_3 is in Q. Since $p_1(1) = q_1(0)$, $p_2(1) = q_2(0)$, it follows that $p_3(1) = kp_1(1) + (1 - k)p_2(1) = kq_1(0) + (1-k)q_2(0) = q_3(0)$, so $p_3 \cdot q_3$ is between $p_1 \cdot q_1$ and $p_2 \cdot q_2$ and hence that $P \cdot Q$ is convex.

14. This can be defined as follows for paths that make turns that are less than 360°: paths $p_1(i)$ and $p_2(i)$ have the same direction if (a) $\varphi_1(0) < \varphi_1(1)$ and $\varphi_2(0) < \varphi_2(1)$ or (b) $\varphi_1(0) > \varphi_1(1)$ and $\varphi_2(0) > \varphi_2(1)$ or (c) $\varphi_1(0) = \varphi_1(1)$ and $\varphi_2(0) = \varphi_2(1)$ and $x_1(0) < x_1(1)$ and $x_2(0) < x_2(1)$ or (d) $\varphi_1(0) = \varphi_1(1)$ and $\varphi_2(0) = \varphi_2(1)$ and $x_1(0) > x_1(1)$ and $x_2(0) > x_2(1)$.

15. For an interesting example of such an analysis, see Ligozat and Condotta (2005).

16. Since "before" and "after" are invariant under multiplication, then if the prepositions were based on the spatial domain and they were invariant under rotation, their corresponding regions would be the full space, as argued earlier.

17. They state that the geometry of the container "may be a primary perceptual indicator of location control" (Garrod et al., 1999, p. 186).

18. "Perturbed" can be given a more precise analysis in terms of minimal changes (δ-changes) of the forces acting on x. Coventry et al. (1994, p. 291) present a similar condition in terms of containment and control, but without using the notion of force.

19. However, this is not always true, as shown by the fact that Swedish uses i (in) for a fly on the ceiling (*flugan i taket*, "the fly on the ceiling").

20. However, "inside the ring" is ambiguous, also meaning inside the tube itself.

21. This measure can be defined for one, two, or three dimensions, depending on the nature of the container.

22. It would be interesting to investigate whether the distinction between the Korean prepositions *kkita* (fitting tightly in) and *netha* (loose inclusion) (Bowerman & Choi, 2001) can be expressed as the difference between whether $in(x) = 0$ or $in(x) > 0$.

23. This example indicates that an analysis of the invariance classes of different prepositions is a strong tool for determining the relevant domains of prepositions.

24. See Tseng and Bergen (2005) for a similar result concerning sign language.

25. In support of forces being basic for these prepositions, Gentner and Bowerman (2009, p. 471) write: "The *op-aan* distinction seems to reflect implicit force dynamics in how the figure (the located object) is related to the ground (the references object). . . . *Op* is used when the figure is viewed as stably in position—not in any salient way acted on by an underlying force that tends to separate it from the ground. . . . *Aan*, in contrast, is used when the figure maintains its position (i.e., resists separation from the ground through forces like gravity or pulling in any direction)."

26. Mathematically, each point on the path should be associated with a force vector (i.e., a force field). The path should therefore be seen as belonging to the force domain and not to the visuospatial domain.

27. This analysis contrasts with that of Lakoff (1987), who analyzes example (11.16) as a separate meaning of "over," generated by a rotational transformation of the central meaning.

28. This can be seen as a special case of the metonymy operation.

29. In Dutch, however, the no-path transformation is not allowed, but in these cases *boven* (above) is obligatory (Beliën, 2008).

30. As noted earlier, spatial uses of "before" may also involve an intentional component.

31. The words "despite," "except," and "regarding," which function as prepositions, seem to depend on the event domain (cf. sec. 3.2.8), in relation to which epistemic relations can be expressed.

12 A Cognitive Analysis of Word Classes

1. Along the same lines, Croft (1990, p. 170) argues that semantic grounding can be given an *iconic* motivation in that "grammatical structure reflects conceptual structure." He writes that "the iconicity hypothesis would propose that the concepts

that fall into the same grammatical category are cognitively similar in some respects" (p. 173).

2. Langacker (2008, p. 112) writes that verbs and nouns are in polar opposition; other word classes are in between.

3. There are also interpretations of "speak" involving the communicative intentions (the illocutionary act) that involve more than the sound domain: "He spoke convincingly."

4. Dixon (2004, p. 26) notes that in some languages, adjectives also have an adverbial function. Such a double function makes sense from the single-domain perspective.

5. This communicative function is different from the use of quantifiers in generics.

6. Cf. Haspelmath's (2003) semantic map for "some/any."

7. This explains why representative-instance quantifiers occur only with count nouns, while proportional quantifiers also occur with mass nouns.

8. Langacker (2003) writes "random selection," but that implies some kind of probabilistic mechanism that I do not think is necessary here.

9. This is reminiscent of Montague's (1974) and Lewis's (1970) type-theoretic construction of quantifiers.

10. Cf. contraction and revisions in the theory of belief change (Gärdenfors, 1988). In contrast, a normal addition to the common ground in a dialogue can be seen as an expansion.

11. A game that is appreciated by many children is to tell them, "Don't think of an elephant!"

12. Although I cannot present any detailed argumentation here, I believe that the principles of language learning that have been discussed here offer a way out of Chomsky's (1980) "poverty of stimulus" argument. Children do not learn the grammar of a language independently of its semantics. The semantic constraints presented in this book will considerably narrow down the possible syntactic systems.

13 Compositionality

1. I use the term "co-text" here, although the basic communicative situation is based on speech.

2. In a totally different spirit, Lewis (1970) uses the compositionality of functions to analyze various linguistic categories.

3. There is a lower bound to decomposition via projection. Either the projection is one-dimensional, so that no further projection makes sense, or it is an integral set of dimensions, that is, a domain. A domain is not decomposable, since the integrality of the dimensions means that it cannot be reconstructed as the product of lower-dimensional projections.

4. In the terminology of Geach (1956), most adjectives are "logically attributive" (while he calls the simple composition "logically predicative").

5. Note that it is not the head but the modifier that is modified before the composition of meanings takes place.

6. Physiologically, what happens is that blood goes away from the surface of the skin. In the fuller color space, this means less red (more gray), but in the small color space, this means moving into the "green" area.

7. See Warglien and Gärdenfors (forthcoming), appendix 2, for a mathematical definition of radial projections.

8. It can be shown that a radial projection establishes a homeomorphism between two convex sets. So long as two sets are convex and compact and have a common interior point, such a homeomorphism always exists (Berge, 1997, p. 167).

9. In Gärdenfors (2000, p. 122), I formulated the following rule for the meaning of the composition of a head D with modifier C: "The combination CD of two concepts C and D is determined by letting the regions for the domains of C, confined to the contrast class defined by D, replace the corresponding regions for D."

10. See Kintsch (2001) for a proposal about how to handle the problems of combinations of nouns and verbs within Latent Semantic Analysis.

14 Modeling Meanings in Robots and in the Semantic Web

1. Searle (1980) argues that they never can, but his argument is based on computation as sequential and based on symbols and on communication merely being an exchange of symbols.

2. Again, I should point out that I do not confine inferences to rule following in some representational system based on symbols.

3. See the discussion of the morning star paradox in section 6.5.3.

4. It is interesting to note that Waismann (1968, p. 128) already argued that "the known relations of logic can only hold between statements which belong to a *homogeneous* domain; or that the deductive nexus never extends beyond the limits of such a domain. Accordingly we may set ourselves the task of arranging the statements of our language in distinct strata."

5. Dietze, Gugliotta, and Domingue (2008) present another approach to handling context in the Semantic Web, also based on conceptual spaces.

6. A slightly related attempt is that of Pustejovsky et al. (2008). They describe GLML—a generative lexicon markup language—that is based on Pustejosky's (1991) theory of the generative lexicon. However, GLML is based on qualia structures and type-theoretical constructions, and it does not consider anything like the geometric structures of domains that form the core of CSML.

15   Taking Stock

1. Yes, I know. The criteria have to some extent been formulated to contrast the proposed theory with other theories. I still believe they have an independent interest.

Appendix: Existence of Fixpoints

1. The text is to a large extent based on Warglien and Gärdenfors (forthcoming).

2. Metrics based on polar coordinates have not been much studied within the cognitive sciences. Given their applicability for analyzing prepositions, as was presented in chapter 11, their general role for cognitive and semantic processes should be further investigated.

3. This follows as a corollary of the classic Bolzano-Weierstrass theorem.

4. Note that I am not modeling communication as such, only its effects on conceptual spaces.

References

Adams, B., & Raubal, M. (2009). The Conceptual Space Markup Language (CSML): Towards the Cognitive Semantic Web. In *Third IEEE International Conference on Semantic Computing* (pp. 253–260). Berkeley, CA: IEEE Computer Society.

Ahlqvist, O. (2004). A parameterized representation of uncertain conceptual spaces. *Transactions in GIS, 8,* 493–514.

Aisbett, J., & Gibbon, G. (1994). A tunable distance measure for coloured solid models. *Artificial Intelligence, 65,* 143–164.

Andersson, R., Holsanova, J., & Holmqvist, K. (submitted). Opinion prediction as communicative success, and the role of visual information in conversation. Manuscript.

Asbury, A., Dotlačil, J., Gehrke, B., & Nouwen, R. (Eds.). (2008). *Syntax and semantics of spatial prepositions.* Amsterdam: John Benjamins.

Astington, J. W., & Jenkins, J. M. (1999). A longitudinal study of the relation between language and theory-of-mind development. *Developmental Psychology, 35,* 1311–1320.

Atkins, B. T., Kegel, J., & Levin, B. (1988). Anatomy of a verb entry: From linguistic theory to lexicographic practice. *International Journal of Lexicography, 12,* 84–126.

Auster, P. (1992). *The New York trilogy: City of glass, Ghosts, The locked room.* London: Faber & Faber.

Bååth, R. (2010). *ChildFreq: An online tool to explore word frequencies in child language.* Lund University Cognitive Studies 16. Lund: Lund University.

Bara, B. G. (2010). *Cognitive pragmatics.* Cambridge, MA: MIT Press.

Barsalou, L. (1999). Perceptual symbol systems. *Behavioral and Brain Sciences, 22,* 577–609.

Bates, E. (Ed.). (1976). *Language and context: The acquisition of pragmatics.* New York: Academic Press.

Bates, E., Camaioni, L., & Volterra, V. (1975). The acquisition of performatives prior to speech. *Merrill-Palmer Quarterly*, *21*, 205–224.

Beliën, M. (2002). Force dynamics in static propositions: Dutch *aan, op*, and *tegen*. In H. Cuyckens & G. Radden (Eds.), *Perspectives on prepositions* (pp. 195–209). Tübingen: Niemeyer.

Beliën, M. (2008). *Constructions, constraints, and construal: Adpositions in Dutch*. Utrecht: LOT Publications.

Ben-Yami, H. (2004). *Logic and natural language: On plural reference and its semantic and logical significance*. Aldershot: Ashgate.

Bennett, J. (1996). What events are. In R. Casati & A. C. Varzi (Eds.), *Events* (pp. 137–151). Aldershot: Dartmouth.

Berge, C. (1997). *Topological spaces*. Mineola, NY: Dover.

Berlin, B., & Kay, P. (1969). *Basic color terms: Their universality and evolution*. Berkeley, CA: University of California Press.

Berners-Lee, T. (1998). What the Semantic Web can represent. Retrieved from http://www.w3.org/DesignIssues/RDFnot.html.

Berners-Lee, T., Hendler, J., & Lassila, O. (2001). The Semantic Web. *Scientific American*, *284*(5), 34–43.

Beyer, O., Camiano, P., & Griffiths, S. (2012). Towards action representation within the framework of conceptual spaces: Preliminary results. *Cognitive Robotics*, AAAI Technical Reports WS-12-06.

Bickerton, D. (2009). *Adam's tongue: How humans made language, how language made humans*. New York: Hill & Wang.

Biederman, I. (1987). Recognition-by-components: A theory of human image understanding. *Psychological Review*, *94*, 115–147.

Bierwisch, M. (1967). Some semantics universals of German adjectivals. *Foundations of Language*, *3*, 1–36.

Billman, D. O. (1983). *Procedures for learning syntactic structure: A model and test with artificial grammars*. Doctoral dissertation, University of Michigan.

Billman, D. O., & Knutson, J. (1996). Unsupervised concept learning and value systematicity: A complex whole aids learning the parts. *Journal of Experimental Psychology: Learning, Memory, and Cognition*, *22*, 458–475.

Blackwell, A. (2000). On the acquisition of the syntax of English adjectives. In A. Okrent & J. P. Boyle (Eds.), *Proceedings from the Panels of the Thirty-Sixth Regional Meeting of the Chicago Linguistic Society* (Vol. 36-2, pp. 361–375). Chicago: Chicago Linguistic Society.

Bleys, J. (2012). Language strategies for color. In L. Steels (Ed.), *Experiments in cultural language evolution* (pp. 61–85). Amsterdam: John Benjamins.

Bloom, P. (2000). *How children learn the meaning of words.* Cambridge, MA: MIT Press.

Bloom, P. (2002). Mindreading, communication, and the learning of names for things. *Mind and Language, 17,* 37–54.

Boesch, C., & Boesch-Achermann, H. (2000). *The chimpanzees of the Tai Forest: Behavioural ecology and evolution.* Oxford: Oxford University Press.

Bohnemeyer, J. (2012). A vector space semantics for reference frames in Yucatec. In E. Bogal-Allbritten (Ed.), *Proceedings of the Sixth Meeting on the Semantics of Under-Represented Languages in the Americas (SULA 6) and SULA-Bar* (pp. 15–34). Amherst: GLSA Publications.

Bowerman, M. (1996a). Learning how to structure space for language: A cross-linguistic perspective. In P. Bloom, M. Peterson, L. Nadel, & M. Garrett (Eds.), *Language and space* (pp. 385–436). Cambridge, MA: MIT Press.

Bowerman, M. (1996b). The origins of children's spatial semantics categories: Cognitive versus linguistic determinants. In J. J. Gumperz & S. C. Levinson (Eds.), *Rethinking linguistic relativity* (pp. 145–176). Cambridge: Cambridge University Press.

Bowerman, M., & Choi, S. (2001). Shaping meanings for language: Universal and language-specific in the acquisition of semantic categories. In M. Bowerman & S. C. Levinson (Eds.), *Language acquisition and conceptual development* (pp. 475–511). Cambridge: Cambridge University Press.

Bowerman, M., & Pedersen, E. (1992). Topological relations picture series. In S. C. Levinson (Ed.), *Space stimuli kit 1.2* (pp. 40–50). Nijmegen: Max Planck Institute for Psycholinguistics.

Brinck, I. (2001). Attention and the evolution of intentional communication. *Pragmatics and Cognition, 9,* 255–272.

Brinck, I. (2004a). The pragmatics of imperative and declarative pointing. *Cognitive Science Quarterly, 3,* 429–446.

Brinck, I. (2004b). Joint attention, triangulation and radical interpretation: A problem and its solution. *Dialectica, 58*(2), 179–205.

Brinck, I. (2008). The role of intersubjectivity for the development of intentional communication. In J. Zlatev, T. Racine, C. Sinha, & E. Itkonen (Eds.), *The shared mind: Perspectives on intersubjectivity* (pp. 115–140). Amsterdam: John Benjamins.

Brinton, L. J. (2000). *The structure of modern English: A linguistic introduction.* Amsterdam: John Benjamins.

Broström, S. (1994). *The role of metaphor in cognitive semantics.* Lund University Cognitive Studies 31. Lund: Lund University.

Brouwer, L. E. J. (1910). Über ein eindeutige, stetige Transformation von Flächen in sich. *Mathematische Annalen, 69,* 176–180.

Browman, C. P., & Goldstein, L. M. (1990). Gestural specification using dynamically-defined articulatory structures. *Journal of Phonetics, 18,* 299–320.

Brugman, C. (1981). *Story of over.* Bloomington, IN: Indiana Linguistics Club.

Bugnyar, T., Stöwe, M., & Heinrich, B. (2004). Ravens, *Corvus corax* follow gaze direction of humans around obstacles. *Proceedings of the Royal Society of London, Series B: Biological Sciences, 271,* 1331–1336.

Bühler, K. (1982). *Sprachtheorie: Die Darstellungsfunktion der Sprache.* Stuttgart: Fischer.

Bullock, D., Cisek, P., & Grossberg, S. (1998). Cortical networks for control of voluntary arm movements under variable force conditions. *Cerebral Cortex, 8,* 48–62.

Butterworth, G., & Jarret, N. L. M. (1991). What minds share in common is space: Spatial mechanisms serving joint visual attention in infancy. *British Journal of Developmental Psychology, 9,* 55–72.

Call, J., & Tomasello, M. (1999). A nonverbal false belief task: The performance of children and great apes. *Child Development, 70,* 381–395.

Camaioni, L., Perucchini, P., Bellagamba, F., & Colonnesi, C. (2004). The role of declarative pointing in developing a theory of mind. *Infancy, 5,* 291–308.

Cangelosi, A., Metta, G., Sagerer, G., Nolfi, S., Nehaniv, C., Fischer, K., et al. (2008). The ITALK project: Integration and transfer of action and language knowledge in robots. In *Proceedings of Third ACM/IEEE International Conference on Human Robot Interaction* (Vol. 2, pp. 167–179). New York: ACM.

Carey, S. (1978). The child as word learner. In M. Halle, J. Bresanan, & G. A. Miller (Eds.), *Linguistic theory and psychological reality* (pp. 264–293). Cambridge, MA: MIT Press.

Carey, S. (1985). *Conceptual change in childhood.* Cambridge, MA: MIT Press.

Carlson, G. N. (2009). Generics and concepts. In J. Pelletier Francis (Ed.), *Kinds, things, and stuff: Mass terms and generics* (pp. 16–36). Oxford: Oxford University Press.

Carlson, L., & van der Zee, E. (2005). *Functional features in language and space: Insights from perception, categorization, and development.* Oxford: Oxford University Press.

Carpenter, M., Nageli, K., & Tomasello, M. (1998). *Social cognition, joint attention, and communicative competence from 9 to 15 months of age.* Monographs of the Society for Research in Child Development, 63(4). New York: Wiley.

Casasanto, D., Fotakopoulou, O., & Boroditsky, L. (2010). Space and time in the child's mind: Evidence for a cross-dimensional asymmetry. *Cognitive Science, 34,* 387–405.

Casati, R., & Varzi, A. (2008). Event concepts. In T. F. Shipley & J. Zacks (Eds.), *Understanding events: From perception to action* (pp. 31–54). New York: Oxford University Press.

Chafe, W. (1995). Accessing the mind through language. In S. Allén (Ed.), *Of thought and words: The relation between language and mind* (pp. 107–125). London: Imperial College Press.

Chella, A., Frixione, M., & Gaglio, S. (2001). Conceptual spaces for computer vision representations. *Artificial Intelligence Review, 16*, 137–152.

Cheney, D., & Seyfarth, R. (1990). *How monkeys see the world: Inside the mind of another species.* Chicago: University of Chicago Press.

Chierchia, G. (1995). *Dynamics of meaning: Anaphora, presupposition, and the theory of grammar.* Chicago: University of Chicago Press.

Chomsky, N. (1980). *Rules and representations.* Oxford: Blackwell.

Chomsky, N. (1986). *Knowledge of language.* New York: Praeger.

Clark, A. (1997). *Being there: Putting brain, body, and world together again.* Cambridge, MA: MIT Press.

Clark, E. V. (1972). On the child's acquisition of antonyms in two semantic fields. *Journal of Verbal Learning and Verbal Behavior, 11*, 750–758.

Clark, E. V. (1973). Non-linguistic strategies and the acquisition of word meanings. *Cognition, 2*, 161–182.

Clark, E. V. (1978). From gesture to word: On the natural history of deixis in language acquisition. In J. S. Bruner & A. Garton (Eds.), *Human growth and development* (pp. 85–120). Oxford: Oxford University Press.

Clark, H. (1992). *Arenas of language use.* Chicago: University of Chicago Press.

Clark, H. (1996). *Using language.* Cambridge: Cambridge University Press.

Clark, H., & Schaefer, E. F. (1989). Contributing to discourse. *Cognitive Science, 13*, 259–294.

Clausner, T. C., & Croft, W. (1999). Domains and image schemas. *Cognitive Linguistics, 10*, 1–31.

Cohen, P. R. (1998). Dynamic maps as representations of verbs. In *Proceedings of the Information and Technology Systems Conference, Fifteenth IFIP World Computer Congress* (pp. 21–33). Dordrecht: Kluwer.

Cohn, A. G., & Renz, J. (2008). Qualitative spatial representation and reasoning. In F. van Hermelen, V. Lifschitz, & B. Porter (Eds.), *Handbook of knowledge representation* (pp. 551–596). Amsterdam: Elsevier.

Cook, R. S., Kay, P., & Regier, T. (2005). The World Color Survey database: History and use. In H. Cohen & C. Lefebvre (Eds.), *Handbook of categorization in cognitive science* (pp. 223–242). Amsterdam: Elsevier.

Corballis, M. C. (2002). *From hand to mouth: The origins of language*. Princeton, NJ: Princeton University Press.

Coseriu, E. (2003). *Geschichte der Sprachphilosophie*. Tübingen, Basel: Francke.

Coventry, K. R., Carmichael, R., & Garrod, S. C. (1994). Spatial prepositions, object-specific function, and task requirements. *Journal of Semantics, 11*, 289–309.

Coventry, K. R., & Garrod, S. C. (2004). *Seeing, saying, and acting: The psychological semantics of spatial prepositions*. Hove and New York: Psychology Press, Taylor & Francis.

Coventry, K. R., Prat-Sala, M., & Richards, L. (2001). The interplay between geometry and function in the comparison of over, under, above, and below. *Journal of Memory and Language, 44*, 376–398.

Croft, W. (1990). *Typology and universals*. Cambridge: Cambridge University Press.

Croft, W. (1991). *Syntactic categories and grammatical relations: The cognitive organization of information*. Chicago: University of Chicago Press.

Croft, W. (1994). The semantics of subjecthood. In M. Yaguello (Ed.), *Subjecthood and subjectivity* (pp. 29–75). Paris: Ophrys.

Croft, W. (2001). *Radical construction grammar: Syntactic theory in typological perspective*. Oxford: Oxford University Press.

Croft, W. (2002). The role of domains in the interpretation of metaphors and metonymies. In R. Dirven & R. Pörings (Eds.), *Metaphor and metonymy in comparison and contrast* (pp. 161–205). Berlin: Mouton de Gruyter.

Croft, W. (2003). *Typology and universals* (2nd ed.). Cambridge: Cambridge University Press.

Croft, W. (2012a). *Verbs: Aspect and argument structure*. Oxford: Oxford University Press.

Croft, W. (2012b). Dimensional models of event structure and verbal semantics. *Theoretical Linguistics, 38*, 195–203.

Croft, W., & Cruse, D. A. (2004). *Cognitive linguistics*. Cambridge: Cambridge University Press.

Croft, W., & Wood, E. J. (2000). Construal operations in linguistics and artificial intelligence. In L. Albertazzi (Ed.), *Meaning and cognition: A multidisciplinary approach* (pp. 51–78). Amsterdam: John Benjamins.

Culicover, P. W., & Jackendoff, R. (2005). *Simpler syntax*. Oxford: Oxford University Press.

Da Silva Sinha, V., Sinha, C., Sampaio, W., & Zinken, J. (2012). Event-based time intervals in an Amazonian culture. To appear in L. Filipović & K. M. Jaszczolt (Eds.), *Space and time across languages and cultures* (Vol. 2): *Language, culture, and cognition* (pp. 15–35). Amsterdam: John Benjamins.

Davidson, D. (1967). The logical form of action sentences. In N. Rescher (Ed.), *The logic of decision and action* (pp. 81–95). Pittsburgh, PA: University of Pittsburgh Press.

Deacon, T. W. (1997). *The symbolic species*. London: Penguin Books.

Deese, J. (1964). The associative structure of some common English adjectives. *Journal of Verbal Learning and Verbal Behavior, 3*, 347–357.

Dekker, P., & van Rooij, R. (2000). Bi-directional optimality theory: An application of game theory. *Journal of Semantics, 17*, 217–242.

DeLancey, S. (1991). Event construal and case role assignment. In *Proceedings of the Seventeenth Annual Meeting of the Berkeley Linguistics Society: General Session and Parasession on the Grammar of Event Structure* (pp. 338–353). Berkeley, CA: Berkeley Linguistics Society.

DeLoache, J. S., Uttal, D. H., & Rosengren, K. S. (2004). Scale errors offer evidence for a perception-action dissociation early in life. *Science, 304*, 1027–1029.

De Lucs, G. (1993). *There are no meaningless sentences*. Lund University Cognitive Studies 16. Lund: Lund University.

Dennett, D. (1991). *Consciousness explained*. Boston, MA: Little, Brown.

D'Entremont, B. (2000). A perceptual-attentional explanation of gaze following in 3- to 6-months-olds. *Developmental Science, 3*, 302–311.

De Saussure, F. (1966). *Cours de linguistique générale*. Paris: Payot.

De Villiers, J., & Pyers, J. (1997). Complementing cognition: The relationship between language and theory of mind. In *Proceedings of the 21st Annual Boston University Conference on Language Development*. Somerville, MA: Cascadilla Press.

Dewell, R. (1994). Over again: Image-schema transformations in semantic analysis. *Cognitive Linguistics, 5*, 351–380.

Dewey, J. (1929). *Experience and nature*. New York: Dover.

Diessel, H. (2006). Demonstratives, joint attention, and the emergence of grammar. *Cognitive Linguistics, 17*, 463–489.

Dietz, R. (2013). Comparative concepts. *Synthese, 190*, 139–170.

Dietze, S., Gugliotta, A., & Domingue, J. (2008). Conceptual situation spaces for situation-driven processes. In *The 5th Annual European Semantic Web Conference.* Galway: Digital Enterprise Research Institute.

Dixon, R. M. W. (1977). Where have all the adjectives gone? *Studies in Language, 1,* 19–80.

Dixon, R. M. W. (2004). Adjective classes in typological perspective. In R. M. W. Dixon & A. Y. Aikhenvald (Eds.), *Adjective classes: A cross-linguistic typology* (pp. 1–49). Oxford: Oxford University Press.

Donald, M. (1991). *Origins of the modern mind.* Cambridge, MA: Harvard University Press.

Douven, I., Decock, L., Dietz, R., & Egré, P. (2011). Vagueness: A conceptual spaces approach. *Journal of Philosophical Logic, 42,* 137–160.

Dowty, D. (1979). *Word meaning and Montague grammar.* Dordrecht: Reidel.

Dowty, D. (1991). Thematic proto-roles and argument selection. *Language, 67,* 547–619.

Dunin-Kepliz, B., & Verbrugge, R. (2001). A tuning machine for cooperative problem solving. *Fundamenta Informatica, 21,* 1001–1025.

Edelberg, W. (2006). Intrasubjective intentional identity. *Journal of Philosophy, 103,* 481–502.

Edelman, S. (1999). *Representation and recognition in vision.* Cambridge, MA: MIT Press.

Emery, N. J. (2000). The eyes have it: The neuroethology, function, and evolution of social gaze. *Neuroscience and Biobehavioral Reviews, 24,* 581–604.

Enfield, N. J. (2003). *Linguistic epidemiology: Semantics and grammar of language contact in mainland Southeast Asia.* London: Routledge Curzon.

Engberg Pedersen, E. (1995). The concept of domain in the cognitive theory of metaphor. *Nordic Journal of Linguistics, 18,* 111–119.

Eschenbach, C., Tschander, L., Habel, C., & Kulik, L. (2000). Lexical specifications of paths. In C. Freksa, W. Brauer, C. Habel, & K. F. Wender (Eds.), *Spatial cognition II: Integrating abstract theories, empirical studies, formal methods, and practical applications* (pp. 127–144). Berlin: Springer.

Evans, V., & Green, M. (2006). *Cognitive linguistics: An introduction.* Edinburgh: Edinburgh University Press.

Fauconnier, G. (1984). *Espaces mentaux: Aspects de la construction du sens dans les langues naturelles.* Paris: Les Éditions de Minuit.

Fauconnier, G. (1990). Invisible meaning. In *Proceedings of the Sixteenth Annual Meeting of the Berkeley Linguistics Society* (pp. 390–404). Berkeley, CA: Berkeley Linguistics Society.

Fauconnier, G., & Turner, R. (1998). Conceptual integration networks. *Cognitive Science, 22*, 133–187.

Feist, M. I., & Gentner, D. (1998). On plates, bowls, and dishes: Factors in the use of English IN and ON. In *Proceedings of the Twentieth Annual Meeting of the Cognitive Science Society* (pp. 345–349). Mahwah, NJ: Erlbaum.

Fenson, L., Dale, P. S., Reznick, J. S., Bates, E., Thal, D., & Pethick, S. (1994). Variability in early communicative development. *Monographs of the Society for Research in Child Development, 59*(5), 1–185.

Fernald, A. (1992). Meaningful melodies in mothers' speech to infants. In H. Papousek, U. Jürgens, & M. Papousek (Eds.), *Nonverbal vocal communication: Comparative and developmental approaches* (pp. 262–282). Cambridge: Cambridge University Press.

Fernández, P. R. (2007). Suppression in metaphor interpretation: Differences between meaning selection and meaning construction. *Journal of Semantics, 24*, 345–371.

Fillmore, C. (1968). The case for case. In E. Bach & E. T. Harms (Eds.), *Universals in linguistic theory* (pp. 1–88). New York: Holt, Rinehart & Winston.

Fillmore, C. (1976). Frame semantics and the nature of language. *Annals of the New York Academy of Sciences: Conference on the Origin and Development of Language and Speech, 280*, 20–32.

Fillmore, C. (1978). On the organization of semantic information in the lexicon. In D. Farkas, W. M. Jacobsen, & K. W. Todrys (Eds.), *Parasession on the lexicon* (pp. 148–173). Chicago: Chicago Linguistics Society.

Fillmore, C. (1982). Frame semantics. In *Linguistics in the morning calm* (pp. 111–137). Seoul: Hanshin Publishing.

Fiorini, S., Gärdenfors, P., & Abel, M. (submitted). Structure, similarity, and spaces. Manuscript.

Fischer Nilsson, J. (1999). A conceptual space logic. In E. Kawaguchi, H. Kangassalo, H. Jaakkola, & I. A. Hamid (Eds.), *Information modelling and knowledge bases XI* (pp. 39–53). Amsterdam: IOS Press.

Flanders, M., Tillery, S., & Soechting, J. (1992). Early stages in a sensorimotor transformation. *Behavioral and Brain Sciences, 15*, 309–320.

Flavell, J. H., Flavell, E. R., Green, F. L., & Moses, J. L. (1990). Young children's understanding of fact beliefs versus value beliefs. *Child Development, 61*, 915–928.

Fodor, J. A. (1983). *The modularity of mind*. Cambridge, MA: MIT Press.

Fodor, J. A., & Katz, J. J. (1963). The structure of a semantic theory. *Language, 39,* 170–210.

Frännhag, H. (2010). *Interpretive Functions of Adjectives in English: A Cognitive Approach*. Doctoral thesis. Lund: Centre for Languages and Literature, Lund University.

Galantucci, B. (2005). An experimental study of the emergence of human communication systems. *Cognitive Science, 29,* 737–767.

Gallant, J. L., Braun, J., & Van Essen, D. C. (1993). Selectivity for polar, hyperbolic, and Cartesian gratings in macaque visual cortex. *Science, 259,* 100–103.

Gallistel, C. R. (1990). *The organization of learning*. Cambridge, MA: MIT Press.

Gärdenfors, P. (1985). Propositional logic based on the dynamics of belief. *Journal of Symbolic Logic, 50,* 390–394.

Gärdenfors, P. (1988). *Knowledge in flux: Modeling the dynamics of epistemic states*. Cambridge, MA: MIT Press.

Gärdenfors, P. (1990). Induction, conceptual spaces, and AI. *Philosophy of Science, 57,* 78–95.

Gärdenfors, P. (1993). The emergence of meaning. *Linguistics and Philosophy, 16,* 285–309.

Gärdenfors, P. (1995). Speaking about the inner environment. In S. Allén (Ed.), *Of thought and words: The relation between language and mind* (pp. 143–151). London: Imperial College Press.

Gärdenfors, P. (1997). Does semantics need reality? In *Does Representation Need Reality?* (pp. 113–120). Austrian Society of Cognitive Science Technical Report 97–01, Vienna.

Gärdenfors, P. (2000). *Conceptual spaces: The geometry of thought*. Cambridge, MA: MIT Press.

Gärdenfors, P. (2003). *How Homo became Sapiens: On the evolution of thinking*. Oxford: Oxford University Press.

Gärdenfors, P. (2004a). Visualizing the meanings of words. In Y. Eriksson & K. Holmqvist (Eds.), *Language and visualization* (pp. 51–69). Lund University Cognitive Studies 119. Lund: Lund University.

Gärdenfors, P. (2004b). How to make the Semantic Web more semantic. In A. C. Varzi & L. Vieu (Eds.), *Formal ontology in information systems* (pp. 19–36). Amsterdam: IOS Press.

Gärdenfors, P. (2004c). Cooperation and the evolution of symbolic communication. In K. Oller & U. Griebel (Eds.), *The evolution of communication systems* (pp. 237–256). Cambridge, MA: MIT Press.

Gärdenfors, P. (2007a). Cognitive semantics and image schemas with embodied forces. In J. M. Krois, M. Rosengren, A. Steidele, & D. Westerkamp (Eds.), *Embodiment in cognition and culture* (pp. 57–76). Amsterdam: Benjamins.

Gärdenfors, P. (2007b). Representing actions and functional properties in conceptual spaces. In T. Ziemke, J. Zlatev, & R. M. Frank (Eds.), *Body, language, and mind* (Vol. 1): *Embodiment* (pp. 167–195). Berlin: Mouton de Gruyter.

Gärdenfors, P. (2007c). Evolutionary and developmental aspect of intersubjectivity. In H. Liljenström & P. Århem (Eds.), *Consciousness transitions: Phylogenetic, ontogenetic, and physiological aspects* (pp. 281–385). Amsterdam: Elsevier.

Gärdenfors, P. (2010). Evolution and semantics. In P. C. Hogan (Ed.), *Cambridge encyclopedia of the language sciences* (pp. 748–750). Cambridge: Cambridge University Press.

Gärdenfors, P., Brinck, I., & Osvath, M. (2012). Coevolution of cooperation, cognition and communication. In F. Stjernfelt, T. Deacon, & T. Schilhab (Eds.), *New perspectives of the symbolic species* (pp. 193–222). Berlin: Springer.

Gärdenfors, P., & Löhndorf, S. (forthcoming). What is a domain? Dimensional structures vs. meronymic relations. *Cognitive Linguistics*.

Gärdenfors, P., & Osvath, M. (2010). Prospection as a cognitive precursor to symbolic communication. In R. Larson, V. Déprez, & H. Yamakido (Eds.), *Evolution of language: Biolinguistic approaches* (pp. 103–114). Cambridge: Cambridge University Press.

Gärdenfors, P., & Warglien, M. (2012). Using conceptual spaces to model actions and events. *Journal of Semantics, 29*, 487–519.

Gärdenfors, P., & Warglien, M. (2013). The development of semantic space for pointing and verbal communication. In J. Hudson, U. Magnusson, & C. Paradis (Eds.), *Conceptual spaces and the construal of spatial meaning: Empirical evidence from human communication* (pp. 29–42). Cambridge: Cambridge University Press.

Gärdenfors P., & Zenker, F. (2013). Theory change as dimensional change: Conceptual spaces applied to the dynamics of empirical theories. *Synthese, 190*, 1039–1058.

Gärdenfors, P., & Zwarts, J. (submitted). Prepositions in conceptual spaces. Manuscript.

Garey, H. B. (1957). Verbal aspect in French. *Language, 33*, 91–110.

Garner, W. R. (1974). *The processing of information and structure.* Potomac, MD: Erlbaum.

Garrod, S., & Anderson, A. (1987). Saying what you mean in dialogue: A study in conceptual and semantic coordination. *Cognition, 27,* 181–218.

Garrod, S. C., Ferrier, G., & Campbell, S. (1999). "In" and "on": Investigating the functional geometry of spatial prepositions. *Cognition, 72,* 167–189.

Geach, P. (1956). Good and evil. *Analysis, 17,* 32–42.

Geach, P. (1967). Intentional identity. *Journal of Philosophy, 64,* 627–632.

Gehrke, B. (2008). *Ps in Motion: On the Semantics and Syntax of P Elements and Motion Events.* Ph.D. dissertation. Utrecht University.

Gennari, S., & Poeppel, D. (2003). Processing correlates of lexical semantics complexity. *Cognition, 89,* B27–B41.

Gentner, D. (1981). Some interesting differences between verbs and nouns. *Cognition and Brain Theory, 4,* 161–178.

Gentner, D., & Boroditsky, L. (2001). Individuation, relativity, and early word learning. In M. Bowerman & S. Levinson (Eds.), *Language acquisition and conceptual development* (pp. 215–256). Cambridge: Cambridge University Press.

Gentner, D., & Bowerman, M. (2009). Why some spatial semantic categories are harder to learn than others: The typological prevalence hypothesis. In J. Guo, E. Lieven, N. Budwig, S. Ervin-Tripp, K. Nakamura, & S. Özçalişkan (Eds.), *Crosslinguistic approaches to the psychology of language: Research in the tradition of Dan Isaac Slobin* (pp. 465–480). New York: Psychology Press.

Gentner, D., & France, I. M. (1988). The verb mutability effect: Studies of the combinatorial semantics of nouns and verbs. In S. L. Small, G. W. Cottrell, & M. K. Tanenhaus (Eds.), *Lexical ambiguity resolution: Perspectives from psycholinguistics, neuropsychology, and artificial intelligence* (pp. 343–382). San Mateo, CA: Kaufman.

Gentner, D., & Kurtz, K. J. (2005). Relational categories. In W. K. Ahn, R. L. Goldstone, B. C. Love, A. B. Markman, & P. W. Wolff (Eds.), *Categorization inside and outside the laboratory* (pp. 151–175). Washington, DC: American Psychological Association.

Gergely, G., & Csibra, G. (2003). Teleological reasoning in infancy: The naive theory of rational action. *Trends in Cognitive Sciences, 7,* 287–292.

Geuder, W., & Weisgerber, M. (2008). Verbs in conceptual space. In G. Katz, S. Reinhart, & P. Reuter (Eds.), *Sinn und Bedeutung VI: Proceeding of the 6th Annual Meeting of the Gesellschaft für Semantik* (pp. 69–84). Osnabrück: Publications of the Institute of Cognitive Science, University of Osnabrück.

Gibbs, R. W., & Colston, H. L. (1995). The cognitive psychological reality of image schemas and their transformations. *Cognitive Linguistics, 6,* 347–378.

Gibson, J. J. (1979). *The ecological approach to visual perception.* Hillsdale, NJ: Erlbaum.

Giese, M. A., & Lappe, M. (2002). Measurement of generalization fields for the recognition of biological motion. *Vision Research, 42,* 1847–1858.

Giese, M. A., & Poggio, T. (2003). Neural mechanisms for the recognition of biological movements. *Nature Reviews: Neuroscience, 4,* 179–192.

Giese, M., Thornton, I., & Edelman, S. (2008). Metrics of the perception of body movement. *Journal of Vision, 8,* 1–18.

Gil, D. (1994). The structure of Riau Indonesian. *Nordic Journal of Linguistics, 17,* 179–200.

Givón, T. (2001). *Syntax* (Vol. 1). Philadelphia, PA: John Benjamins.

Goddard, C., & Wierzbicka, A. (1994). *Semantic and lexical universals: Theory and empirical findings.* Amsterdam: John Benjamins.

Goldberg, A. (1995). *Constructions: A construction grammar approach to argument structure.* Chicago: University of Chicago Press.

Goldberg, A. (2010). Verbs, construction and semantic frames. In M. Rappaport Hovav, E. Doron, & I. Sichel (Eds.), *Syntax, lexical semantics, and event structure* (pp. 39–58). Oxford: Oxford University Press.

Goldin-Meadow, S. (2007). Pointing sets the stage for learning language—and creating language. *Child Development, 78,* 741–745.

Goldstone, R. L. (1994). The role of similarity in categorization: Providing a groundwork. *Cognition, 52,* 125–157.

Goldstone, R. L., & Barsalou, L. (1998). Reuniting perception and conception. *Cognition, 65,* 231–262.

Gómez, J. C. (1994). Mutual awareness in primate communication: A Gricean approach. In S. T. Parker, R. W. Mitchell, & M. L. Boccia (Eds.), *Self-awareness in animals and humans* (pp. 61–80). Cambridge: Cambridge University Press.

Gómez, J. C. (2007). Pointing behaviors in apes and human infants: A balanced interpretation. *Child Development, 78,* 729–734.

Goodall, J. (1986). *The chimpanzees of Gombe: Patterns of behavior.* Cambridge, MA: Harvard University Press.

Gopnik, A., & Astington, J. W. (1988). Children's understanding of representational change, and its relation to the understanding of false belief and the appearance-reality distinction. *Child Development, 59,* 26–37.

Gopnik, A., & Meltzoff, A. N. (1997). *Words, thoughts, and theories*. Cambridge, MA: MIT Press.

Gräfenhain, M., Behne, T., Carpenter, M., & Tomasello, M. (2009). Young children's understanding of joint commitments. *Developmental Psychology, 45,* 1430–1443.

Grice, P. (1975). Logic and conversation. In P. Cole & J. L. Morgan (Eds.), *Syntax and semantics* (Vol. 3): *Speech acts* (pp. 41–58). New York: Academic Press.

Gruber, J. S. (1967). Look and see. *Language, 43,* 937–947.

Guizzardi, G. (2005). *Ontological foundations for structural conceptual models*. Enschede: CTIT.

Habel, C. (1989). Zwischen-Bericht. In C. Habel, M. Herweg, & K. Rehkämper (Eds.), *Raumkonzepte in Verstehensprozessen* (pp. 37–69). Tübingen: Niemeyer.

Harder, P. (2010). *Meaning in mind and society: A functional contribution to the social turn in cognitive linguistics*. Berlin: De Gruyter Mouton.

Harnad, S. (1987). Category induction and representation. In S. Harnad (Ed.), *Categorical perception* (pp. 535–565). Cambridge: Cambridge University Press.

Harnad, S. (1990). The symbol grounding problem. *Physica D: Nonlinear Phenomena, 42,* 335–346.

Harnad, S. (2005). To cognize is to categorize: Cognition is categorization. In C. Lefebvre & H. Cohen (Eds.), *Handbook of categorization in cognitive science* (pp. 20–42). New York: Elsevier.

Harner, L. (1981). Children talk about the time and aspect of actions. *Child Development, 52,* 498–506.

Haspelmath, M. (1997). *Indefinite pronouns*. Oxford: Oxford University Press.

Haspelmath, M. (2003). The geometry of grammatical meaning: Semantic maps and cross-linguistic comparison. In M. Tomasello (Ed.), *The new psychology of language* (Vol. 2, pp. 211–242). Mahwah, NJ: Erlbaum.

Hasson, U., Ghazanfar, A. A., Galantucci, B., Garrod, S., & Keysers, C. (2012). Brain-to-brain coupling: A mechanism for creating and sharing a social world. *Trends in Cognitive Sciences, 16,* 114–121.

Haviland, J. B. (2000). Pointing, gesture space and mental maps. In D. McNeill (Ed.), *Language and gesture* (pp. 13–46). Cambridge: Cambridge University Press.

Heim, I. (1982). *The semantics of definite and indefinite noun phrases*. Doctoral dissertation. Amherst, MA: University of Massachusetts.

Hemeren, P. E. (1996). Frequency, ordinal position and semantic distance as measures of cross-cultural stability and hierarchies for action verbs. *Acta Psychologica, 91,* 39–66.

Hemeren, P. E. (1997). Typicality and context effects in action categories. In *Proceedings of the 19th Annual Conference of the Cognitive Science Society* (p. 949). Stanford, CA: Erlbaum.

Hemeren, P. E. (2008). *Mind in action*. Lund University Cognitive Studies 140. Lund: Lund University.

Herskovits, A. (1986). *Language and spatial cognition: An interdisciplinary study of the prepositions in English*. Cambridge: Cambridge University Press.

Hidaka, S., & Smith, L. B. (2011). Packing: A geometric analysis of feature selection and category formation. *Cognitive Systems Research, 12*, 1–18.

Hilton, C. E., & Meldrum, D. J. (2004). Striders, runners, and transporters. In D. J. Meldrum & C. E. Hilton (Eds.), *From biped to strider: The emergence of modern human walking, running, and resource transport* (pp. 1–8). New York: Kluwer Academic.

Hockett, C. F. (1960). The origin of speech. *Scientific American, 203*(3), 88–96.

Holland, J. H., Holyoak, K. J., Nisbett, R. E., & Thagard, P. R. (1986). *Induction: Processes of inference, learning, and discovery*. Cambridge, MA: MIT Press.

Holmqvist, K. (1993). *Implementing cognitive semantics*. Lund University Cognitive Studies 17. Lund: Lund University.

Holsanova, J. (2008). *Discourse, vision, and cognition*. Amsterdam: John Benjamins.

Holyoak, K. J., & Thagard, P. (1996). *Mental leaps*. Cambridge, MA: MIT Press.

Honingh, A., & Bod, R. (2011). In search of universal properties of musical scales. *Journal of New Music Research, 40*, 81–89.

Hundsnurscher, F., & Splett, J. (1982). *Semantik der Adjektive im Deutschen: Analyze der semantischen Relationen*. Wiesbaden: Westdeutscher.

Hunn, E. (1976). A measure of the degree of correspondence of folk to scientific biological classification. *American Ethnologist, 2*, 309–327.

Hurford, J. (1999). The evolution of language and languages. In R. Dunbar, C. Knight, & C. Power (Eds.), *The evolution of culture* (pp. 173–193). Edinburgh: Edinburgh University Press.

Hurford, J. R. (2007). *The origins of meaning: Language in the light of evolution*. Oxford: Oxford University Press.

Huth, A. G., Nishimoto, S., Wu, A. T., & Gallant, J. L. (2012). A continuous semantic space describes the representation of thousands of object and action categories across the human brain. *Neuron, 76*, 1210–1224.

Huttenlocher, J., Smiley, P., & Charney, R. (1983). Emergence of actions categories in the child: Evidence from verb meanings. *Psychological Review, 90*, 72–93.

Iverson, J., & Goldin-Meadow, S. (2005). Gesture paves the way for language development. *Psychological Science, 16*, 367–371.

Jackendoff, R. (1972). *Semantic interpretation in generative grammar.* Cambridge, MA: MIT Press.

Jackendoff, R. (1976). Toward an explanatory semantic representation. *Linguistic Inquiry, 7*, 89–150.

Jackendoff, R. (1983). *Semantics and cognition.* Cambridge, MA: MIT Press.

Jackendoff, R. (1985). Multiple subcategorization and the theta-criterion: The case of climb. *Natural Language and Linguistic Theory, 3*, 271–295.

Jackendoff, R. (1987a). *Consciousness and the computational mind.* Cambridge, MA: MIT Press.

Jackendoff, R. (1987b). On beyond zebra: The relation of linguistic and visual information. *Cognition, 26*, 80–114.

Jackendoff, R. (1987c). The status of thematic relations in linguistic theory. *Linguistic Inquiry, 18*, 369–411.

Jackendoff, R. (1990). *Semantic structures.* Cambridge, MA: MIT Press.

Jackendoff, R. (1991). Parts and boundaries. *Cognition, 41*, 9–45.

Jackendoff, R. (1994). *Patterns in the mind: Language and human nature.* New York: Basic Books.

Jackendoff, R. (2007). *Language, consciousness, and culture: Essays on mental structure.* Cambridge, MA: MIT Press.

Jackendoff, R. (2012). Language as a source of evidence for theories of spatial representation. *Perception, 41*, 1128–1152.

Jackendoff, R., & Aaron, D. (1991). Review of G. Lakoff and M. Turner, *More than cool reason: A field guide to metaphor. Language, 67*, 320–338.

Jacot, J. (2012). Do we speak of the same witch? How minds can meet on intentional identity. Lund: Department of Philosophy, Lund University.

Jäger, G. (2007). The evolution of convex categories. *Linguistics and Philosophy, 30*, 551–564.

Jäger, G. (2010). Natural color categories are convex sets. *Amsterdam Colloquium 2009, LNAI 6042* (pp. 11–20). Berlin: Springer.

Jäger, G., Metzger, L. P., & Riedel, F. (2011). Equilibria in cheap-talk games with high-dimensional types and few signals. *Games and Economic Behavior, 73*, 517–537.

Jäger, G., & van Rooij, R. (2007). Language structure: Psychological and social constraints. *Synthese*, *159*, 99–130.

James, W. (1890). *The principles of psychology*. New York: Holt.

Jammer, M. (1957). *Concepts of force: A study in the foundations of dynamics*. Cambridge, MA: Harvard University Press.

Johannesson, M. (2002). *Geometric models of similarity*. Lund University Cognitive Studies 87. Lund: Lund University.

Johansson, G. (1973). Visual perception of biological motion and a model for its analysis. *Perception and Psychophysics*, *14*, 201–211.

Johansson, R., Holsanova, J., & Holmqvist, K. (2006). Pictures and spoken descriptions elicit similar eye movements during mental imagery, both in light and in complete darkness. *Cognitive Science*, *30*, 1053–1079.

Johnson, M. (1987). *The body in the mind: The bodily basis of cognition*. Chicago: University of Chicago Press.

Kalkan, S., Dag, N., Yürüten, O., Borghi, A. M., & Sahin, E. (forthcoming). Verb concepts from affordances. Manuscript.

Kaminski, J., Riedel, J., Call, J., & Tomasello, M. (2005). Domestic goats, *Capra hircus*, follow gaze direction and use social cues in an object task. *Animal Behaviour*, *69*, 11–18.

Kamp, H. (1981). A theory of truth and semantic representation. In J. A. G. Groenendijk, T. M. V. Janssen, & M. B. J. Stokhof (Eds.), *Formal methods in the study of language* (pp. 277–322). Mathematical Centre Tracts 135. Amsterdam: Mathematisch Centrum.

Kaplan, F. (2000). *L'émergence d'un lexique dans une population d'agents autonomes*. Paris: Laboratoire d'Informatique de Paris 6.

Katz, J. J. (1966). *The philosophy of language*. New York: Harper & Row.

Kaufmann, I. (1995). What is an (im-)possible verb? Restrictions on semantic form and their consequences for argument structure. *Folia Linguistica*, *29*, 67–103.

Keenan, E. J. (1984). Semantic correlates of the ergative/absolutive distinction. *Linguistics*, *22*, 197–223.

Kemler Nelson, D. G. (1993). Processing integral dimensions: The whole view. *Journal of Experimental Psychology: Human Perception and Performance*, *19*, 1105–1113.

Kemp, C., & Regier, T. (2012). Kinship categories across languages reflect general communicative principles. *Science*, *336*, 1049–1054.

Kim, J. (1976). Events as property exemplifications. In M. Brand & D. Walton (Eds.), *Action theory* (pp. 159–177). Dordrecht: Reidel.

Kintsch, W. (2001). Predication. *Cognitive Science, 25,* 173–201.

Kiparsky, P. (1997). Remarks on denominal verbs. In A. Alsina, J. Bresnan, & P. Sells (Eds.), *Complex predicates* (pp. 473–499). Stanford, CA: Center for the Study of Language and Information.

Kirby, S. (1999). *Function, selection, and innateness: The emergence of language universals.* Oxford: Oxford University Press.

Kirby, S., & Hurford, J. (2002). The emergence of linguistic structure: An overview of the iterated learning model. In A. Cangelosi & D. Parisi (Eds.), *Simulating the evolution of language* (pp. 121–148). London: Springer.

Koontz-Garboden, A., & Beavers, J. (2012). Manner and result in the roots of verbal meaning. *Linguistic Inquiry, 43,* 331–369.

Kosslyn, S. M. (1980). *Image and mind.* Cambridge, MA: Harvard University Press.

Kosslyn, S. M. (1996). *Image and brain: The resolution of the imagery debate.* Cambridge, MA: MIT Press.

Kracht, M., & Klein, U. (2012). Against the single-domain constraint. *Theoretical Linguistics, 38,* 211–221.

Krauss, R. M., & Glucksberg, S. (1977). Social and non-social speech. *Scientific American, 236,* 100–105.

Kreitzer, A. (1997). Multiple levels of schematization: A study in the conceptualization of space. *Cognitive Linguistics, 8,* 291–325.

Krifka, M. (2012a). Some remarks on event structure, conceptual spaces, and the semantics of verbs. *Theoretical Linguistics, 38,* 223–236.

Krifka, M. (2012b). Definitional generics. In A. Mari, C. Beyssade, & F. Prete (Eds.), *Genericity* (pp. 372–389). Oxford: Oxford University Press.

Kruskal, J. B. (1964). Multidimensional scaling by optimizing goodness of fit to a nonmetric hypothesis. *Psychometrika, 29,* 1–27.

Labov, W. (1973). The boundaries of words and their meanings. In J. Fishman (Ed.), *New ways of analyzing variation in English* (pp. 340–373). Washington, DC: Georgetown University Press.

Lakoff, G. (1987). *Women, fire, and dangerous things.* Chicago: University of Chicago Press.

Lakoff, G. (1990). The invariance hypothesis. *Cognitive Linguistics, 1,* 39–74.

Lakoff, G. (1993). The contemporary theory of metaphor. In A. Ortony (Ed.), *Metaphor and thought* (2nd ed., pp. 202–251). Cambridge: Cambridge University Press.

Lakoff, G., & Johnson, M. (1980). *Metaphors we live by*. Chicago: University of Chicago Press.

Lakoff, G., & Turner, M. (1989). *More than cool reason: A field guide to poetic metaphor*. Chicago: University of Chicago Press.

Lakoff, R. (1971). IFs, ANDs, and BUTs: about conjunction. In C. Fillmore & D. T. Langendoen (Eds.), *Studies in linguistic semantics* (pp. 114–149). New York: Holt, Rinehart & Winston.

Lallee, S., Madden, C., Hoen, M., & Dominey, P. F. (2010). Linking language with embodied and teleological representations of action for humanoid cognition. *Frontiers in Neurobotics*, *4*, 1–12.

Landau, B., & Jackendoff, R. (1993). "What" and "where" in spatial language and spatial cognition. *Behavioral and Brain Sciences*, *16*, 217–265.

Landau, B., Smith, L., & Jones, S. (1998). Object perception and object naming in early development. *Trends in Cognitive Sciences*, *2*, 19–24.

Langacker, R. W. (1986a). An introduction to cognitive grammar. *Cognitive Science*, *10*, 1–40.

Langacker, R. W. (1986b). Settings, participants, and grammatical relations. In *Proceedings of the Annual Meeting of the Pacific Linguistics Conference* (Vol. 2, pp. 1–31). Eugene, OR: University of Oregon.

Langacker, R. W. (1987). *Foundations of cognitive grammar* (Vol. 1). Stanford, CA: Stanford University Press.

Langacker, R. W. (1991a). *Foundations of cognitive grammar* (Vol. 2): *Descriptive application*. Stanford: Stanford University Press.

Langacker, R. W. (1991b). *Concept, image, symbol*. Berlin: Mouton de Gruyter.

Langacker, R. W. (2003). One any. In *Korean Linguistics Today and Tomorrow: Proceedings of the 2002 International Conference on Korean Linguistics* (pp. 282–300). Seoul: Association for Korean Linguistics.

Langacker, R. W. (2008). *Cognitive grammar: A basic introduction*. Oxford: Oxford University Press.

Langley, P. (1996). *Elements of machine learning*. San Francisco, CA: Morgan Kaufmann.

Larsson, B. (1997). *Le bon sens commun: Remarques sur le rôle de la (re)cognition intersubjective dans l'épistémologie et l'ontologie du sens*. Lund: Lund University Press.

Larsson, B. (2008). Le sens commun ou la sémantique comme science de intersubjectivité humaine. *Langages, 170*, 28–40.

Lawler, J. M. (1973). Studies in English generics. University of Michigan Papers in Linguistics. Ann Arbor, MI: University of Michigan.

Leavens, D. A., Hopkins, W. D., & Bard, K. A. (2005). Understanding the point of chimpanzee pointing: Epigenesis and ecological validity. *Current Directions in Psychological Science, 14*, 185–189.

Leavens, D. A., Hopkins, W. D., & Bard, K. A. (2008). The heterochronic origins of explicit reference. In J. Zlatev, T. Racine, C. Sinha, & E. Itkonen (Eds.), *The shared mind: Perspectives on intersubjectivity* (pp. 187–214). Amsterdam: John Benjamins.

Lee, I., & Portier, B. (2007). An empirical study of knowledge representation and learning within conceptual spaces for intelligent agents. In *6th IEEE/ACIS International Conference on Computer and Information Science* (pp. 463–468). Los Alamitos, CA: IEEE Computer Society.

Lee, S. M. (1994). *Untersuchungen zur Valenz des Adjektivs in der deutschen Gegenwartssprache*. Bern: Peter Lang.

Leibniz, G. W. (1765). Nouveaux essais sur l'entendement humain. In *Philosophische Schriften* (Vol. 6, pp. 39–527). Berlin: Akademie.

Leslie, A. M. (1994). ToMM, ToBy, and Agency: Core architecture and domain specificity. In L. A. Hirschfeld & S. A. Gelman (Eds.), *Mapping the mind: Domain specificity in cognition and culture* (pp. 119–148). Cambridge: Cambridge University Press.

Levin, B. (1993). *English verb classes and alternations*. Chicago: University of Chicago Press.

Levin, B., & Rappaport Hovav, M. (1991). Wiping the slate clean: A lexical semantics exploration. *Cognition, 41*, 123–151.

Levin, B., & Rappaport Hovav, M. (2005). *Argument realization*. Cambridge: Cambridge University Press.

Levin, B., & Rappaport Hovav, M. (forthcoming). Lexicalised meaning and manner/result complementarity. In B. Arsenijevic, B. Gehrke, & R. Marin (Eds.), *Subatomic semantics of event predicates*. Dordrecht: Springer.

Levinson, S. C. (1996). Frames of reference and Malyneux's question: Cross-linguistic evidence. In P. Bloom, M. A. Peterson, L. Nadel, & M. Garret (Eds.), *Language and space* (pp. 109–170). Cambridge, MA: MIT Press.

Levinson, S. C. (2003). *Space in language and cognition: Explorations in cognitive diversity*. Cambridge: Cambridge University Press.

Lewis, D. K. (1969). *Convention*. Cambridge, MA: Harvard University Press.

Lewis, D. K. (1970). General semantics. *Synthese, 22,* 18–67.

Lewis, D. K. (1979). Scorekeeping in a language game. *Journal of Philosophical Logic,* 8, 339–359.

Ligozat, G., & Condotta, J.-F. (2005). On the relevance of conceptual spaces for spatial and temporal reasoning. *Spatial Cognition and Computation, 5,* 1–27.

Liszkowski, U., Brown, P., Callaghan, T., Takada, A., & de Vos, C. (2011). A Prelinguistic gestural universal of human communication. *Cognitive Science, 39,* 698–713.

Liszkowski, U., Carpenter, M., Henning, A., Striano, T., & Tomasello, M. (2004). Twelve-month-olds point to share attention and interest. *Developmental Science, 7,* 297–307.

Liszkowski, U., Carpenter, M., & Tomasello, M. (2007). Reference and attitude in infant pointing. *Journal of Child Language, 34,* 1–20.

Liszkowski, U., Schäfer, M., Carpenter, M., & Tomasello, M. (2009). Prelinguistic infants, but not chimpanzees, communicate about absent entities. *Psychological Science, 20,* 654–660.

Locke, J. [1690] (1959). *An essay concerning human understanding.* New York: Dover.

Lupyan, G., & Dale, R. (2010). Language structure is partly determined by social structure. *PLoS ONE, 5*(1), e8559. doi:10.1371/journal.pone.0008559.

Lyons, J. (1977). *Semantics.* Cambridge: Cambridge University Press.

Macnamara, J. (1978). How can we talk about what we see? Department of Psychology, McGill University.

MacWhinney, B. (1987). *Mechanisms of language acquisition.* Hillsdale, NJ: Erlbaum.

Maddox, W. T. (1992). Perceptual and decisional separability. In G. F. Ashby (Ed.), *multidimensional models of perception and cognition* (pp. 147–180). Hillsdale, NJ: Erlbaum.

Majid, A., Gullberg, M., van Staden, M., & Bowermann, M. (2007). How similar are semantic categories in closely related languages? A comparison of cutting and breaking in four Germanic languages. *Cognitive Linguistics, 18,* 179–194.

Maling, J. (1983). Transitive adjectives: A case of categorical reanalysis. In F. Heny & B. Richards (Eds.), *Linguistic categories: Auxiliaries and related puzzles* (Vol. 1, pp. 253–289). Dordrecht: Reidel.

Malt, B., Ameel, E., Imai, M., Gennari, S., Saji, N. M., & Majid, A. (submitted). Human locomotion in languages: Constraints on moving and meaning. Manuscript.

Mandler, J. M. (2004). *The foundations of mind: Origins of conceptual thought.* New York: Oxford University Press.

Mani, I., & Pustejovsky, J. (2012). *Interpreting motion in language.* Oxford: Oxford University Press.

Markman, E. M. (1994). Constraints on word meaning in early language acquisition. *Lingua, 92,* 199–227.

Markman, E. M. (1992). Constraints on word learning: Speculations about their nature, origins, and domain specificity. In M. R. Gunnar & M. P. Maratsos (Eds.), *Modularity and constraints in language and cognition: The Minnesota Symposia on Child Psychology* (pp. 59–101). Hillsdale, NJ: Erlbaum.

Marr, D., & Nishihara, K. H. (1978). Representation and recognition of the spatial organization of three-dimensional shapes. *Proceedings of the Royal Society of London, Series B: Biological Sciences, 200,* 269–294.

Marr, D., & Vaina, L. (1982). Representation and recognition of the movements of shapes. *Proceedings of the Royal Society of London, Series B: Biological Sciences, 214,* 501–524.

Matlock, T. (2004). Fictive motion as cognitive simulation. *Memory and Cognition, 32,* 1389–1400.

McDonough, L. (2002). Basic-level nouns: First learned but misunderstood. *Journal of Child Language, 29,* 357–377.

McIntyre, A. (2007). Functional interpretations: Borderline idiosyncrasy in prepositional phrases and other expressions. Retrieved from http://www2.unine.ch/files/content/sites/andrew.mcintyre/files/shared/mcintyre/functional.pdf.

McKay, T. J. (2006). *Plural predication.* Oxford: Clarendon Press.

McNeill, D. (1992). *Hand and mind: What gestures reveal about thought.* Chicago: University of Chicago Press.

Mead, G. H. (1934). *Mind, self, and society.* Chicago: University of Chicago Press.

Meinong, A. (1960). On the theory of objects. In R. Chisholm (Ed.), *Realism and the background of phenomenology* (pp. 76–117). Glencoe, IL: Free Press.

Melara, R. D. (1992). The concept of perceptual similarity: From psychophysics to cognitive psychology. In D. Algom (Ed.), *Psychophysical approaches to cognition* (pp. 303–388). Amsterdam: Elsevier.

Mervis, C., & Rosch, E. (1981). Categorization of natural objects. *Annual Review of Psychology, 32,* 89–115.

Michotte, A. (1963). *The perception of causality.* London: Methuen.

Miller, G. A., & Johnson-Laird, P. N. (1976). *Language and perception*. Cambridge, MA: Belknap Press.

Mintz, T. B., & Gleitman, L. R. (2002). Adjectives really do modify nouns: The incremental and restricted nature of early adjective acquisition. *Cognition, 84,* 267–293.

Mitchell, P. (1997). *Introduction to theory of mind: Children, autism, and apes*. London: Arnold.

Moltmann, F. (1998). Part structures, integrity, and the mass-count distinction. *Synthese, 116,* 75–111.

Montague, R. (1970). Universal grammar. *Theoria, 36,* 373–398.

Montague, R. (1974). *Formal philosophy*. R. H. Thomason (Ed.). New Haven, CT: Yale University Press.

Nelson, K. (1996). *Language in cognitive development*. Cambridge: Cambridge University Press.

Nelson, K. (2005). Language pathways into the community of minds. In J. W. Astington & J. A. Baird (Eds.), *Why language matters for theory of mind* (pp. 26–49). Oxford: Oxford University Press.

Newell, A., & Simon, H. (1972). *Human problem solving*. Englewood Cliffs, NJ: Prentice-Hall.

Nikitina, T. (2008). Pragmatic factors and variation in the expression of spatial goals: The case of *into* vs. *in*. In A. Asbury, J. Dotlačil, B. Gehrke, & R. Nouwen (Eds.), *Syntax and semantics of spatial prepositions* (pp. 175–195). Amsterdam: John Benjamins.

Nosofsky, R. M. (1986). Attention, similarity, and the identification-categorization relationship. *Journal of Experimental Psychology: General, 115,* 39–57.

Nosofsky, R. M. (1988). Similarity, frequency, and category representations. *Journal of Experimental Psychology: Learning, Memory, and Cognition, 14,* 54–65.

Nosofsky, R. M. (1992). Similarity scaling and cognitive process models. *Annual Review of Psychology, 43,* 25–53.

Noy, N. F., & McGuinness, D. L. (2001). Ontology development 101: A guide to creating your first ontology. Stanford Knowledge Systems Laboratory Technical Report, Stanford, CA.

Ojeda, A. (1993). *Linguistic individuals*. Stanford, CA: Center for the Study of Language and Information.

Okabe, A., Boots, B., & Sugihara, K. (1992). *Spatial tessellations: Concepts and applications of Voronoi diagrams*. New York: John Wiley & Sons.

O'Keefe, J. (1996). The spatial prepositions in English, vector grammar, and the cognitive map theory. In P. Bloom, M. A. Peterson, L. Nadel, & M. F. Garrett (Eds.), *Language and space* (pp. 277–316). Cambridge, MA: MIT Press.

Onishi, K. H., & Baillargeon, R. (2005). Fifteen-month-old infants understand false beliefs. *Science, 308,* 255–257.

Osgood, C. E., Suci, G. J., & Tannenbaum, P. (1957). *The measurement of meaning.* Urbana: University of Illinois Press.

Pantcheva, M. (2010). The syntactic structure of locations, goals, and sources. *Linguistics, 48,* 1043–1081.

Paradis, C. (2001). Adjectives and boundedness. *Cognitive Linguistics, 12,* 47–65.

Paradis, C. (2005). Ontologies and construals in lexical semantics. *Axiomathes, 15,* 541–573.

Paradis, C. (2008). Configurations, construals, and change: Expressions of DEGREE. *English Language and Linguistics, 12,* 317–343.

Parikh, P. (2010). *Language and equilibrium.* Cambridge, MA: MIT Press.

Parsons, T. (1980). *Nonexistent objects.* New Haven, CT: Yale University Press.

Parsons, T. (1990). *Events in the semantics of English.* Cambridge, MA: MIT Press.

Peirsman, Y., & Geeraerts, D. (2006). Metonymy as a prototypical category. *Cognitive Linguistics, 17,* 269–316.

Pentland, A. P. (1986). Perceptual organization and the representation of natural form. *Artificial Intelligence, 28,* 293–331.

Perlmutter, D. M., & Postal, P. M. (1984). The l-advancement exclusiveness law. In D. M. Perlmutter & C. Rosen (Eds.), *Studies in relational grammar* (pp. 81–125). Chicago: Chicago University Press.

Perner, J., Leekam, S., & Wimmer, H. (1987). Three-year-old's difficulty with false belief: The case for a conceptual deficit. *British Journal of Developmental Psychology, 5,* 125–137.

Petitot, J. (1989). Morphodynamics and the categorical perception of phonological units. *Theoretical Linguistics, 15,* 25–71.

Petitot, J. (2011). *Cognitive morphodynamics.* Bern: Peter Lang.

Piaget, J. (1927/1969). *The child's conception of time.* Harmondworth: Penguin.

Piaget, J. (1954). *The construction of reality in the child.* New York: Basic Books.

Piaget, J. (1972). *The psychology of the child.* New York: Basic Books.

Pickering, M. J., & Garrod, S. (2004). Toward a mechanistic psychology of dialogue. *Behavioral and Brain Sciences, 27,* 169–190.

Pinker, S. (1989). *Learnability and cognition: The acquisition of argument structure.* Cambridge, MA: MIT Press.

Polunin, O. (2009). Temporal dimension of the framing effect in topical mental accounting. *Studia Psychologica, 51,* 343–355.

Port, R. F., & van Gelder, T. (Eds.). (1995). *Mind as motion.* Cambridge, MA: MIT Press.

Premack, D., & Woodruff, G. (1978). Does the chimpanzee have a theory of mind? *Behavioral and Brain Sciences, 4,* 515–526.

Preston, S. D., & de Waal, F. (2003). Empathy: Its ultimate and proximal bases. *Behavioral and Brain Sciences, 25,* 1–72.

Preuschoft, H., & Witte, H. (1991). Biomechanical reasons for the evolution of hominid body shape. In Y. Coppens & B. Senut (Eds.), *Origine de la bipedie chez les hominides* (pp. 55–79). Paris: Cahiers de Paleoanthropologie Editions du CNRS.

Pulvermüller, F. (2003). *Neuroscience of language: On brain circuits of words and serial order.* Cambridge: Cambridge University Press.

Pustejovsky, J. (1991). The generative lexicon. *Computational Linguistics, 17,* 409–441.

Pustejovsky, J. (1995). A survey of dot objects. Technical Report. Brandeis University.

Pustejovsky, J. (2001). Type construction and the logic of concepts. In P. Bouillon & F. Busa (Eds.), *The language of word meaning* (pp. 91–123). Cambridge: Cambridge University Press.

Pustejovsky, J., Rumshisky, A., Moszkowicz, J. L., & Batuikova, O. (2008). GLML: A Generative Lexicon Markup Language. Retrieved from http://glml-italian.wdfiles .com/local--files/home/glml.pdf.

Putnam, H. (1975). The meaning of "meaning." In K. Gunderson (Ed.), *Language, mind, and knowledge* (pp. 131–193). Minneapolis: University of Minnesota Press.

Quine, W. V. O. (1960). *Word and object.* Cambridge, MA: MIT Press.

Rachidi, R. (1989). *Gegensatzrelationen im Bereich deutscher Adjective.* Tübingen: Max Niemeyer.

Rappaport Hovav, M., & Levin, B. (1998). Building verb meanings. In M. Butt & W. Geuder (Eds.), *The projection of arguments: Lexical and compositional factors* (pp. 97–134). Stanford, CA: Center for the Study of Language and Information.

Rappaport Hovav, M., & Levin, B. (2002). Change of state verbs: Implications for theories of argument projection. In *Proceedings of the 28th Annual Meeting of the Berkeley Linguistics Society* (pp. 269–280). Berkeley, CA: Berkeley Linguistics Society.

Rappaport Hovav, M., & Levin, B. (2010). Reflections on manner/result complementarity. In M. Rappaport Hovav, D. Doron, & I. Sichel (Eds.), *Lexical semantics, syntax, and event structure* (pp. 21–38). Oxford: Oxford University Press.

Raubal, M. (2004). Formalizing conceptual spaces. In A. C. Varzi & L. Vieu (Eds.), *Formal ontology in information systems* (pp. 153–164). Amsterdam: IOS Press.

Ravid, D. (2006). Semantic development in textual contexts during the school years: Noun scale analyses. *Journal of Child Language, 33,* 791–821.

Reddy, M. J. (1979). The conduit metaphor: A case of frame conflict in our language about language. In A. Ortony (Ed.), *Metaphor and thought* (pp. 284–310). Cambridge: Cambridge University Press.

Reed, S. K. (1972). Pattern recognition and categorization. *Cognitive Psychology, 3,* 382–407.

Regier, T. (1996). *The human semantic potential: Spatial language and constrained connectionism.* Cambridge, MA: MIT Press.

Regier, T., Khetarpal, N., & Majid, A. (forthcoming). Inferring semantic maps. To appear in *Linguistic Typology.*

Reichenbach, H. (1928). *Philosophie der Raum-Zeit-Lehre.* Berlin: Walter de Gruyter.

Repacholi, B., & Gopnik, A. (1997). Early understanding of desires: Evidence from 14 and 18 month olds. *Developmental Psychology, 33,* 12–21.

Richardson, D. C., & Dale, R. (2005). Looking to understand: The coupling between speakers' and listeners' eye movements and its relationship to discourse comprehension. *Cognitive Science, 29,* 1045–1060.

Rosch, E. (1975). Cognitive representations of semantic categories. *Journal of Experimental Psychology: General, 104,* 192–233.

Rosch, E. (1978). Prototype classification and logical classification: The two systems. In E. Scholnik (Ed.), *New trends in cognitive representation: Challenges to Piaget's theory* (pp. 73–86). Hillsdale, NJ: Erlbaum.

Rosen, S. T. (1999). The syntactic representation of linguistic events. *Glot International, 4,* 3–11.

Runesson, S. (1994). Perception of biological motion: The KSD-principle and the implications of a distal versus proximal approach. In G. Jansson, S.-S. Bergström, & W. Epstein (Eds.), *Perceiving events and objects* (pp. 383–405). Hillsdale, NJ: Erlbaum.

Runesson, S., & Frykholm, G. (1981). Visual perception of lifted weights. *Journal of Experimental Psychology: Human Perception and Performance*, *7*, 733–740.

Runesson, S., & Frykholm, G. (1983). Kinematic specification of dynamics as an informational basis for person and action perception: Expectation, gender recognition, and deceptive intention. *Journal of Experimental Psychology: General*, *112*, 585–615.

Russell, J. A. (1980). A circumplex model of affect. *Journal of Personality and Social Psychology*, *39*, 1161–1178.

Saint-Dizier, P. (2006). *The linguistic dimension of prepositions and their use in NLP applications*. Dordrecht: Kluwer.

Sanso, A. (2010). How conceptual are semantics maps? *Linguistic Discovery*, *8*, 288–309.

Schank, R. C. (1975). *Conceptual information processing*. New York: Elsevier Science.

Schank, R. C., & Abelson, R. P. (1977). *Scripts, plans, goals, and understanding: An inquiry into human knowledge structures*. Hillsdale, NJ: Erlbaum.

Schelling, T. (1960). *The strategy of conflict*. Cambridge, MA: Harvard University Press.

Schiffman, H. R. (1982). *Sensation and perception* (2nd ed.). New York: Wiley.

Schwarzschild, R. (2008). The semantics of comparatives and other degree constructions. *Language and Linguistics Compass*, *2*, 308–311.

Schwering, A., & Raubal, M. (2005). Measuring semantic similarity between geospatial conceptual regions. In M. A. Rodriguez, F. Cruz, M. J. Egenhofer, & S. Levashkin (Eds.), *Geospatial Semantics 2005* (pp. 90–106). Heidelberg: Springer.

Searle, J. (1969). *Speech acts: An essay in the philosophy of language*. Cambridge: Cambridge University Press.

Searle, J. (1980). Minds, brains, and programs. *Behavioral and Brain Sciences*, *3*, 417–457.

Searle, J. (1983). *Intentionality: An essay in the philosophy of mind*. Cambridge: Cambridge University Press.

Seston, R., Michnick Golinkoff, R., Weyiy, M., & Hirsh-Pasek, K. (2009). Vacuuming with my mouth? Children's ability to comprehend novel extensions of familiar verbs. *Cognitive Development*, *24*, 113–124.

Shepard, R. N. (1987). Toward a universal law of generalization for psychological science. *Science*, *237*, 1317–1323.

Shirky, C. (2003). The Semantic Web, syllogism, and worldview. http://www.shirky.com/writings/semantic_syllogism.html.

Simon, H. (1969). *The sciences of the artificial*. Cambridge, MA: MIT Press.

Simons, P. (1991). *Parts: A study in ontology*. Oxford: Clarendon Press.

Sinha, C., Da Silva Sinha, V., Zinken, J., & Sampaio, W. (2011). When time is not space: The social and linguistic construction of time intervals and temporal event relations in an Amazonian culture. *Language and Cognition, 3*, 137–169.

Sivik, L., & Taft, C. (1994). Color naming: A mapping in the NCS of common color terms. *Scandinavian Journal of Psychology, 35*, 144–164.

Smiley, P., & Huttenlocher, J. (1995). Conceptual development and the child's early words for events, objects, and persons. In M. Tomasello & W. E. Merriman (Eds.), *Beyond names for things: Young children's acquisition of verbs* (pp. 21–61). Hillsdale, NJ: Erlbaum.

Smith, E. E., & Medin, D. L. (1981). *Categories and concepts*. Cambridge, MA: Harvard University Press.

Smith, L. B. (1989). From global similarities to kinds of similarities—the construction of dimensions in development. In S. Vosniadou & A. Ortony (Eds.), *Similarity and analogical reasoning* (pp. 146–178). Cambridge: Cambridge University Press.

Smith, L. B. (2005). Action alters shape categories. *Cognitive Science, 29*, 665–679.

Smith, L. B. (2009). From fragments to geometric shape: Changes in visual object recognition between 18 and 24 months. *Current Directions in Psychological Science, 18*, 290–294.

Smith, L. B., & Heise, D. (1992). Perceptual similarity and conceptual structure. In B. Burns (Ed.), *Percepts, concepts, and categories* (pp. 231–272). Amsterdam: Elsevier.

Smith, L. B., & Samuelson, L. (2006). An attentional learning account of the shape bias. *Developmental Psychology, 42*, 1339–1343.

Smith, L. B., & Sera, M. D. (1992). A developmental analysis of the polar structure of dimensions. *Cognitive Psychology, 24*, 99–142.

Solt, S. (2011). How many most's? In I. Reich, E. Horch, & D. Pauly (Eds.), *Proceedings of Sinn und Bedeutung 15*, 565–579. Saarbrücken: Universaar—Saarland University Press.

Son, J. Y., Smith, L. B., & Goldstone, R. L. (2008). Simplicity and generalization: Short-cutting abstraction in children's object categorizations. *Cognition, 108*, 626–638.

Southgate, V., Senju, A., & Csibra, G. (2007). Action anticipation through attribution of false belief by 2-year-olds. *Psychological Science, 16*, 587–592.

Spelke, E. S. (1994). Initial knowledge: Six suggestions. *Cognition, 50*, 431–445.

Spelke, E. S. (2000). Core knowledge. *American Psychologist* (November): 1233–1243.

Spelke, E. S., & Kinzler, K. D. (2007). Core knowledge. *Developmental Science, 10,* 89–96.

Sperber, D., & Wilson, D. (2002). Pragmatics, modularity, and mind-reading. *Mind and Language, 17,* 3–23.

Stalnaker, R. (1978). Assertion. *Syntax and Semantics, 9,* 315–332.

Stalnaker, R. (2002). Common ground. *Linguistics and Philosophy, 25,* 701–721.

Steels, L. (1999). *The talking heads experiment.* Antwerp: Laboratorium.

Steels, L., & Belpaeme, T. (2005). Coordinating perceptually grounded categories through language: A case study for colour. *Behavioral and Brain Sciences, 28,* 469–489.

Steels, L., & Kaplan, F. (2002). Bootstrapping grounded word semantics. In T. Briscoe (Ed.), *Linguistic evolution through language acquisition: Formal and computational models* (pp. 53–74). Cambridge: Cambridge University Press.

Stern, D. N. (1985). *The interpersonal world of the infant: A view from psychoanalysis and developmental psychology.* New York: Basic Books.

Svenonius, P. (2006). The emergence of axial parts. *Nordlyd, 33,* 49–77.

Svenonius, P. (2012). Structural decomposition of spatial adpositions. Paper presented at The Meaning of P, Bochum, November 24, 2012.

Sweetser, E. (1990). *From etymology to pragmatics.* Cambridge: Cambridge University Press.

Szabo, Z. (2004). Compositionality. Retrieved from http://plato.stanford.edu/entries/compositionality.

Taft, C., & Sivik, L. (1997). Salient color terms in four languages. *Scandinavian Journal of Psychology, 38,* 26–31.

Talmy, L. (1972). Semantic structures in English and Atsugewi. Berkeley, CA: Department of Linguistics, University of California.

Talmy, L. (1975). Semantics and syntax of motion. In J. P. Kimball (Ed.), *Syntax and semantics* (Vol. 4, pp. 181–238). New York: Academic Press.

Talmy, L. (1976). Semantic causative types: The grammar of causative constructions. In M. Shibatani (Ed.), *Syntax and semantics* (Vol. 6, pp. 43–116). New York: Academic Press.

Talmy, L. (1985). Lexicalization patterns: Semantic structure in lexical forms. In T. Shopen (Ed.), *Language typology and syntactic description 3: Grammatical categories and the lexicon* (pp. 57–149). Cambridge: Cambridge University Press.

Talmy, L. (1988). Force dynamics in language and cognition. *Cognitive Science, 12,* 49–100.

Talmy, L. (1996). Fictive motion in language and "ception." In P. Bloom, M. A. Peterson, L. Nadel, & M. F. Garrett (Eds.), *Language and space* (pp. 211–276). Cambridge, MA: MIT Press.

Talmy, L. (2000). *Toward a cognitive semantics* (Vol. 1): *Concept structuring systems.* Cambridge, MA: MIT Press.

Talmy, L. (2001). *Toward a cognitive semantics* (Vol. 2): *Typology and process in concept structuring.* Cambridge, MA: MIT Press.

Talmy, L. (2011). Semantics, universals of. In *Cambridge encyclopedia of language* (pp. 754–757). Cambridge: Cambridge University Press.

Taylor, J. R. (1989). *Linguistic categorization: Prototypes in linguistic theory.* Oxford: Oxford University Press.

Thelen, E., & Smith, L. B. (1994). *A dynamic systems approach to the development of cognition and action.* Cambridge, MA: MIT Press.

Thom, R. (1970). Topologie et linguistique. In A. Haefliger & R. Narasimhan (Eds.), *Essays on topology and related topics* (pp. 226–248). Berlin: Springer.

Thomason, S. K. (1989). Free construction of time from events. *Journal of Philosophical Logic, 18,* 43–67.

Tikhanoff, V., Cangelosi, A., & Metta, G. (2011). Integration of speech and action in humanoid robots: iCub simulation experiments. *IEEE Transactions on Autonomous Mental Development, 3,* 17–29.

Tomasello, M. (1999). *The cultural origins of human cognition.* Cambridge, MA: Harvard University Press.

Tomasello, M. (2001). Perceiving intentions and learning words in the second year of life. In M. Bowermann & S. Levinson (Eds.), *Language acquisition and conceptual development* (pp. 135–158). Cambridge: Cambridge University Press.

Tomasello, M., & Call, J. (2006). Do chimpanzees know what others see—or only what they are looking at? In S. Hurley & M. Nudds (Eds.), *Rational animals* (pp. 371–384). Oxford: Oxford University Press.

Tomasello, M., Carpenter, M., Call, J., Behne, T., & Moll, H. (2005). Understanding and sharing intentions: The origins of cultural cognition. *Behavioral and Brain Sciences, 28,* 675–735.

Tomasello, M., Carpenter, M., & Liszkowski, U. (2007). A new look at infant pointing. *Child Development, 78,* 705–722.

Tourangeau, R., & Sternberg, R. J. (1982). Understanding and appreciating metaphors. *Cognition, 11*, 203–244.

Traugott, E. C. (2012). *Pragmatics and language change: The Cambridge handbook of pragmatics* (pp. 549–566). Cambridge: Cambridge University Press.

Traugott, E. C., & Dasher, R. B. (2002). *Regularity in semantic change*. Cambridge: Cambridge University Press.

Tribushinina, E., & Gillis, S. (2012). The acquisition of scalar structures: Production of adjectives and degree markers by Dutch-speaking children and their caregivers. *Linguistics, 50*, 241–268.

Troje, N. F. (2008). Retrieving information from human movement patterns. In T. F. Shipley & J. M. Zacks (Eds.), *Understanding events: How humans see, represent, and act on events* (pp. 308–334). Oxford: Oxford University Press.

Tseng, M. J., & Bergen, B. K. (2005). Lexical processing drives motor simulation. In *Proceedings of the Twenty-seventh Annual Conference of the Cognitive Science Society* (pp. 2206–2211). Mahwah, NJ: Erlbaum.

Tversky, B., & Hemenway, K. (1984). Objects, parts, and categories. *Journal of Experimental Psychology: General, 113*, 169–191.

Tylén, K., Weed, E., Wallentin, M., Roepstoorf, A., & Frith, C. D. (2010). Language as a tool for interacting minds. *Mind and Language, 25*, 3–29.

Tyler, A., & Evans, V. (2001). Reconsidering prepositional polysemy networks: The case of *over*. *Language, 77*, 724–765.

Vaina, L. (1983). From shapes and movements to objects and actions. *Synthese, 54*, 3–36.

Vaina, L., & Bennour, Y. (1985). A computational approach to visual recognition of arm movement. *Perceptual and Motor Skills, 60*, 203–228.

Van Benthem, J. (2008). Games that make sense: Logic, language, and multi-agent interaction. In K. Apt & R. van Rooij (Eds.), *New perspectives on games and interaction* (pp. 197–209). Amsterdam: Amsterdam University Press.

Van Dam, W. O., Rueschemeyer, S.-A., & Bekkering, H. (2010). How specifically are action verbs represented in the neural motor system: An fMRI study. *NeuroImage, 53*, 1308–1325.

Vandeloise, C. (1986). *L'espace en français: Sémantique des prépositions spatiales*. Paris: Editions du Seuil.

Van der Gucht, F., Klaas, W., & De Cuypere, L. (2007). The iconicity of embodied meaning: Polysemy of spatial prepositions in the cognitive framework. *Language Sciences, 29*, 733–754.

Van der Zee, E., & Watson, M. (2004). Between space and function: How spatial and functional features determine the comprehension of between. In L. Carlson & E. van der Zee (Eds.), *Functional features in language and space: Insights from perception, categorization and development* (pp. 116–127). Oxford: Oxford University Press.

Van Gelder, T. (1995). What might cognition be, if not computation? *Journal of Philosophy*, *92*, 345–381.

Varzi, A. (2003). Mereology. In E. N. Zalta (Ed.), *The Stanford encyclopedia of philosophy*, http://plato.stanford.edu/archives/win2012/entries/mereology.

Vendler, Z. (1957). Verbs and times. *Philosophical Review*, *56*, 97–121.

Verhagen, A. (2005). *Constructions of intersubjectivity: Discourse, syntax, and cognition*. Oxford: Oxford University Press.

Viberg, Å. (2010). Basic verbs of possession: A contrastive and typological study. *CogniTextes*, *4*, retrieved from http://cognitextes.revues.org/308.

Visser, U. (2004). *Intelligent information integration for the Semantic Web*. Berlin: Springer.

Vulchanova, M., & van der Zee, E. (2012). *Motion encoding in language and space*. Oxford: Oxford University Press.

Vygotsky, L. S. (1978). *Mind in society*. Cambridge, MA: Harvard University Press.

Wagner, L., & Lakusta, L. (2009). Using language to navigate the infant mind. *Perspectives on Psychological Science*, *4*, 177–184.

Waismann, F. (1968). Verifiability. In A. G. N. Flew (Ed.), *Logic and language* (pp. 117–144). Oxford: Blackwell.

Wang, W., Crompton, R. H., Carey, T. S., Günther, M. M., Li, Y., Savage, R., et al. (2004). Comparison of inverse-dynamics musculo-skeletal models of AL 288–1 *Australopithecus afarensis* and KNM-WT 15000 *Homo ergaster* to modern humans, with implications for the evolution of bipedalism. *Journal of Human Evolution*, *47*, 453–478.

Warglien, M. (2001). Playing conversation games. Paper presented at the 2001 Wittgenstein Society Symposium, Kirchberg.

Warglien, M., & Gärdenfors, P. (2013). Semantics, conceptual spaces, and the meeting of minds. *Synthese*, *190*, 2165–2193.

Warglien, M., Gärdenfors, P., & Westera, M. (2012). Event structure, conceptual spaces, and the semantics of verbs. *Theoretical Linguistics*, *38*, 159–193.

Waxman, S. R., & Markow, D. B. (1998). Object properties and object kind: Twenty-one-month-old infants' extension of novel adjectives. *Child Development*, *69*, 1313–1329.

Wellens, P., & Loetzsch, M. (2012). Multi-dimensional meanings in lexicon formation. In L. Steels (Ed.), *Experiments in cultural language evolution* (pp. 143–166). Amsterdam: John Benjamins.

Wellman, H. M., & Liu, D. (2004). Scaling of theory-of-mind tasks. *Child Development*, *75*, 523–541.

Westera, M. (2008). *Action representations and the semantics of verbs*. Bachelor's thesis. Cognitive Artificial Intelligence, Utrecht University.

White, P. A. (1995). *The understanding of causation and the production of action*. Hove: Erlbaum.

Williams, M.-A., McCarthy, J., Gärdenfors, P., Stanton, C., & Karol, A. (2009). A grounding framework. *Journal of Autonomous Agents and Multi-agent Systems*, *19*, 272–296.

Winter, S. (1998). *Expectations and linguistic meaning*. Lund University Cognitive Studies 71. Lund: Lund University.

Winter, S., & Gärdenfors, P. (1995). Linguistic modality as expressions of social power. *Nordic Journal of Linguistics*, *18*, 137–166.

Wisniewski, E. J. (1996). Construal and similarity in conceptual combination. *Journal of Memory and Language*, *35*, 434–453.

Wittgenstein, L. (1953). *Philosophical investigations*. New York: Macmillan.

Wolf, P., & Shepard, J. (2013). Causation, touch, and the perception of force. In B. H. Ross (Ed.), *The psychology of learning and motivation* (pp. 167–202). New York: Academic Press.

Wolff, P. (2007). Representing causation. *Journal of Experimental Psychology: General*, *136*, 82–111.

Wolff, P. (2008). Dynamics and the perception of causal events. In T. Shipley & J. Zacks (Eds.), *Understanding events: How humans see, represent, and act on events* (pp. 555–587). Oxford: Oxford University Press.

Wolff, P. (2012). Representing verbs with force vectors. *Theoretical Linguistics*, *38*, 237–248.

Woodward, A. (1998). Infants selectively encode the goal object of an actor's reach. *Cognition*, *69*, 1–34.

Wunderlich, D. (1991). How do prepositional phrases fit into compositional syntax and semantics. *Linguistics*, *29*, 591–621.

Yu, C., & Smith, L. B. (2012). Embodied attention and word learning by toddlers. Department of Psychological and Brain Sciences Cognitive Science Program, Indiana University.

Zhu, S. C., & Yuille, A. L. (1996). FORMS: A flexible object recognition and modelling system. *International Journal of Computer Vision, 20,* 187–212.

Zlatev, J. (1997). *Situated embodiment: Studies in the emergence of spatial meaning.* Stockholm: Gotab.

Zlatev, J. (2003). Polysemy or generality? Mu. In H. Cuyckens, R. Dirven, & J. R. Taylor (Eds.), *Cognitive approaches to lexical semantics* (pp. 447–494). Berlin: Mouton de Gruyter.

Zlatev, J., Persson, T., & Gärdenfors, P. (2005). *Bodily mimesis as "the missing link" in human cognitive evolution.* Lund University Cognitive Studies 121. Lund: Lund University.

Zwarts, J. (2003). Paths round a prototype. In P. Saint-Dizier (Ed.), *The linguistic dimensions of prepositions and their use in computational formalisms and applications* (pp. 228–238). Toulouse: IRIT.

Zwarts, J. (2005). Prepositional aspect and the algebra of paths. *Linguistics and Philosophy, 28,* 739–779.

Zwarts, J. (2006). Event shape: Paths in the semantics of verbs. Manuscript, Radboud University Nijmegen and Utrecht University.

Zwarts, J. (2008). Aspects of a typology of direction. In S. D. Rothstein (Ed.), *Theoretical and crosslinguistic approaches to the semantics of aspect* (pp. 79–106). Amsterdam: John Benjamins.

Zwarts, J. (2010a). Forceful prepositions. In V. Evans & P. Chilton (Eds.), *Language, cognition, and space: The state of the art and new directions* (pp. 193–214). London: Equinox.

Zwarts, J. (2010b). Semantic map geometry: Two approaches. *Linguistic Discovery, 8,* 377–395.

Zwarts, J., & Winter, Y. (2000). Vector space semantics: A model-theoretic analysis of locative prepositions. *Journal of Logic Language and Information, 9,* 171–213.

Index